Rudolf Holze
Experimental Electrochemistry

D1471767

Related Titles

Hamann, C. H., Hamnett, A.,
Vielstich, W.

Electrochemistry

2007
ISBN: 978-3-527-31069-2

Bard, A. J., Stratmann, M., Gileadi, E.,
Urbakh, M., Calvo, E. J., Unwin, P. R.,
Frankel, G. S., Macdonald, D., Licht, S.,
Schäfer, H. J., Wilson, G. S., Rubinstein, I.,
Fujihira, M., Schmuki, P., Scholz, F.,
Pickett, C. J., Rusling, J. F. (eds.)

**Encyclopedia
of Electrochemistry**
11 Volume Set

2007
ISBN: 978-3-527-30250-5

Wang, J.

Analytical Electrochemistry

2006
ISBN: 978-0-471-67879-3

Savéant, J.-M.

**Elements of Molecular
and Biomolecular
Electrochemistry**

2006
ISBN: 978-0-471-44573-9

Bagotsky, V. S. (ed.)

**Fundamentals
of Electrochemistry**

2005
ISBN: 978-0-471-70058-6

Bard, A. J.

**Electrochemical Methods –
Fundamentals & Applications
2e Student Solutions
Manual (WSE)**

2002
ISBN: 978-0-471-40521-4

Bard, A. J., Faulkner, L. R.

Electrochemical Methods
Fundamentals and Applications

2001
ISBN: 978-0-471-04372-0

Hodes, G. (ed.)

**Electrochemistry
of Nanomaterials**

2001
ISBN: 978-3-527-29836-5

Rudolf Holze

Experimental Electrochemistry

A Laboratory Textbook

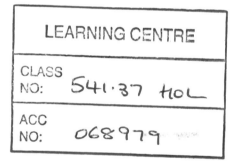

WILEY-VCH

WILEY-VCH Verlag GmbH & Co. KGaA

The Author

Prof. Dr. Rudolf Holze
Chemnitz University of Technology
Institute of Chemistry
Straße der Nationen 62
09111 Chemnitz
Germany

Library of Congress Card No.: applied for

British Library Cataloguing-in-Publication Data
A catalogue record for this book is available from the British Library.

**Bibliographic information published
by the Deutsche Nationalbibliothek**
Die Deutsche Nationalbibliothek lists this publication in the Deutsche Nationalbibliografie; detailed bibliographic data are available in the Internet at http://dnb.d-nb.de.

Composition   K+V Fotosatz GmbH, Beerfelden
Printing   Betz-Druck GmbH, Darmstadt
Bookbinding   Litges & Dopf Buchbinderei GmbH, Heppenheim

Printed in the Federal Republic of Germany
Printed on acid-free paper

ISBN   978-3-527-31098-2

Contents

Preface *IX*

Foreword *XIII*

Symbols and Acronyms *XV*

1 Introduction – An Overview of Practical Electrochemistry *1*
Practical Hints *3*
Electrodes *3*
Measuring Instruments *7*
Electrochemical Cells *8*
Data Recording *10*

2 Electrochemistry in Equilibrium *11*
Experiment 2.1: The Electrochemical Series *11*
Experiment 2.2: Standard Electrode Potentials
and the Mean Activity Coefficient *15*
Experiment 2.3: pH-Measurements and Potentiometrically
Indicated Titrations *21*
Experiment 2.4: Redox Titrations (Cerimetry) *26*
Experiment 2.5: Differential Potentiometric Titration *28*
Experiment 2.6: Potentiometric Measurement of the Kinetics
of the Oxidation of Oxalic Acid *32*
Experiment 2.7: Polarization and Decomposition Voltage *36*

3 Electrochemistry with Flowing Current *43*
Experiment 3.1: Ion Movement in an Electric Field *44*
Experiment 3.2: Paper Electrophoresis *46*
Experiment 3.3: Charge Transport in Electrolyte Solution *47*
Experiment 3.4: Conductance Titration *52*
Experiment 3.5: Chemical Constitution and Electrolytic Conductance *54*

Experimental Electrochemistry. A Laboratory Textbook. Rudolf Holze
Copyright © 2009 WILEY-VCH Verlag GmbH & Co. KGaA, Weinheim
ISBN: 978-3-527-31098-2

Experiment 3.6: Faraday's Law   *56*
Experiment 3.7: Kinetics of Ester Saponification   *59*
Experiment 3.8: Movement of Ions and Hittorf Transport Number   *63*
Experiment 3.9: Polarographic Investigation of the Electroreduction
of Formaldehyde   *69*
Experiment 3.10: Galvanostatic Measurement
of Stationary Current-Potential Curves   *74*
Experiment 3.11: Cyclic Voltammetry   *77*
Experiment 3.12: Slow Scan Cyclic Voltammetry   *85*
Experiment 3.13: Kinetic Investigations with Cyclic Voltammetry   *88*
Experiment 3.14: Numerical Simulation of Cyclic Voltammograms   *93*
Experiment 3.15: Cyclic Voltammetry with Microelectrodes   *95*
Experiment 3.16: Cyclic Voltammetry of Organic Molecules   *99*
Experiment 3.17: Cyclic Voltammetry in Nonaqueous Solutions   *105*
Experiment 3.18: Cyclic Voltammetry
with Sequential Electrode Pocesses   *107*
Experiment 3.19: Cyclic Voltammetry of Aromatic Hydrocarbons   *110*
Experiment 3.20: Cyclic Voltammetry of Aniline and Polyaniline   *113*
Experiment 3.21: Galvanostatic Step Measurements   *118*
Experiment 3.22: Chronoamperometry   *122*
Experiment 3.23: Chronocoulometry   *124*
Experiment 3.24: Rotating Disc Electrode   *126*
Experiment 3.25: Rotating Ring-Disc Electrode   *131*
Experiment 3.26: Measurement of Electrode Impedances   *134*
Experiment 3.27: Corrosion Cells   *137*
Experiment 3.28: Aeration Cell   *139*
Experiment 3.29: Concentration Cell   *141*
Experiment 3.30: Salt Water Drop Experiment According to Evans   *142*
Experiment 3.31: Passivation and Activation of an Iron Surface   *143*
Experiment 3.32: Cyclic Voltammetry with Corroding Electrodes   *145*
Experiment 3.33: Oscillating Reactions   *147*

4 Analytical Electrochemistry   *151*
Experiment 4.1: Ion-sensitive Electrode   *152*
Experiment 4.2: Potentiometrically Indicated Titrations   *154*
Experiment 4.3: Bipotentiometrically Indicated Titration   *159*
Experiment 4.4: Conductometrically Indicated Titration   *161*
Experiment 4.5: Electrogravimetry   *163*
Experiment 4.6: Coulometric Titration   *166*
Experiment 4.7: Amperometry   *168*
Experiment 4.8: Polarography (Fundamentals)   *174*
Experiment 4.9: Polarography (Advanced Methods)   *178*
Experiment 4.10: Anodic Stripping Voltammetry   *180*
Experiment 4.11: Abrasive Stripping Voltammetry   *183*

Experiment 4.12: Polarographic Analysis of Anions   *185*
Experiment 4.13: Tensammetry   *188*

5 Non-Traditional Electrochemistry   *195*
Experiment 5.1: UV-Vis Spectroscopy   *195*
Experiment 5.2: Surface Enhanced Raman Spectroscopy   *199*
Experiment 5.3: Infrared Spectroelectrochemistry   *201*
Experiment 5.4: Electrochromism   *203*

6 Electrochemical Energy Conversion and Storage   *205*
Experiment 6.1: Lead Acid Accumulator   *205*
Experiment 6.2: Discharge Behavior of Nickel-Cadmium Accumulators   *210*
Experiment 6.3: Performance Data of a Fuel Cell   *213*

7 Electrochemical Production   *217*
Experiment 7.1: Cementation Reaction   *217*
Experiment 7.2: Galvanic Copper Deposition   *218*
Experiment 7.3: Electrochemical Oxidation of Aluminum   *220*
Experiment 7.4: Kolbe Electrolysis of Acetic Acid   *222*
Experiment 7.5: Electrolysis of Acetyl Acetone   *223*
Experiment 7.6: Anodic Oxidation of Malonic Acid Diethylester   *226*
Experiment 7.7: Indirect Anodic Dimerization
of Acetoacetic Ester (3-oxo-butyric acid ethyl ester)   *227*
Experiment 7.8: Electrochemical Bromination of Acetone   *229*
Experiment 7.9: Electrochemical Iodination of Ethano   *231*
Experiment 7.10: Electrochemical Production
of Potassium Peroxodisulfate   *233*
Experiment 7.11: Yield of Chlor-alkali Electrolysis According
to the Diaphragm Process   *234*

Appendix   *237*

Index   *239*

Preface

Electrochemistry, taught as a subject at all levels from advanced classes in high school to the research of PhD students, is an extremely interdisciplinary science. Electrochemical processes, methods, models, and concepts are present in numerous fields of science and technology. This clearly illustrates the extremely interdisciplinary character of this branch of science. Accordingly, the points of contact with this science are numerous at all levels of education. Being an experimental science, electrochemistry demands the personal experience – the direct hands-on test of a model or a theory is more convincing than anything else. Consequently, at all levels of education, electrochemical experiments of different degrees of complexity are to be found. The intensity of the interaction ranges from the simple application of an electrochemical instrument (e.g., in a pH measurement or the electrolytic generation of hydrogen) up to complete electrochemical laboratory courses as offered in many universities. The increasing importance of the numerous applications of electrochemistry in sensors, surface technology, materials science, microsystems technology, and nanotechnology will certainly help to enhance this importance and omnipresence.

After publication in 1953 of the ninth and final edition of the book "Elektrochemisches Praktikum", initially published in 1931 by Erich Müller, no textbook in German providing a collection of descriptions of reproducible electrochemical experiments illuminating the whole scope of this science has been published. English textbooks are only slightly more recent. The textbook by N.J. Selley: Experimental Approach to Electrochemistry (Edward Arnold, London 1977) has been out of print for some time. The workbook by J. O'M. Bockris and R.A. Fredlein: A Workbook of Electrochemistry (Plenum Press, New York – London 1973), being a useful supplement from the theoretical point of view though without any experiments described in it, has met the same fate. Thus there seems to be a considerable need for a textbook containing a collection of descriptions of reproducible experiments suitable for course work at all levels from advanced high school to graduate school at university. The Eurocurriculum Electrochemistry developed by the Federation of European Chemical Societies as approved in 1999 called (so far without success) for descriptions of laboratory exercises supplementing the numerous already available textbooks. This curriculum was taken as a guideline. The selection presented here is based on an extensive collection of experiments developed and installed as part of laboratory courses for students of chemistry, materials science, and other sciences also. In addition it contains experiments developed for teachers at various levels in school where pupils will encounter electrochemistry for the first time. Because the whole range of electrochemistry can hardly be present as a whole at a single place and in one research group, experiments and their descriptions as supplied by instructors from other universities are included. Thus, special thanks are due to F. Beck, H. Schäfer, J.-W. Schultze, M. Paul, K. Banert, H.J. Thomas, R. Daniel Little and E. Steckhan[†]. The development of the described experiments and the corresponding instructions in the author's group would have been impossible without the enthusiastic cooperation of creative students and researchers. W. Leyffer, K. Pflugbeil, J. Poppe, and M. Stelter have provided invaluable support by careful evaluation and optimization of experimental concepts; numerous students in laboratory courses have provided results and further input; this is gratefully acknowledged. Finally, E. Rahm has tested many descriptions for practical applications; without her, many minor and perhaps even some major defects would have made it into print. Part of the manuscript was prepared during a stay at St. Petersburg State University: the generous hospitality of my host V. Malev and the stimulating environment as well as a travel stipend of DAAD are gratefully acknowledged.

The scope of electrochemistry is not only illustrated by the diversity of methods and concepts; it is also demonstrated by the range of instruments and tools employed. The experiments described here range from simple tests easily performed at school to complex investigations requiring spectrometers and other large instruments most likely feasible only in a university laboratory. Thus, the author hopes to provide some stimulating input for teachers at both limits of the range: the high school teacher looking for an experiment demonstrating ionization as well as the university professor extending his physical or organic chemistry laboratory course.

In all descriptions emphasis is placed on clear, well-defined, and lucid descriptions, including all details needed for successful repetition of the experiment. Unnecessary details are avoided. Practical details of an experimental setup and instruction for manufacturing are added only if really necessary. Safety instructions and suggestions for safe experiments are provided only in case of particular dangers. Complete lists of dangers, risks, and safety instructions, which are very likely present in every chemistry laboratory are not given. Instrument manufactures are not mentioned or suggested as this might cause undesirable reluctance in installing a given experiment. Only specific characteristics of an instrument required for, e.g., a spectroscopic experiment, are stated.

The present book cannot supplement a textbook of electrochemistry. Any attempt would have resulted in a book of excessive size, which in addition would be hard to use. Instead, brief introductions and some background are provided at the beginning of every description, supplemented by references to the textbook by C.H. Hamann, A. Hamnett, and W. Vielstich: Electrochemistry, Second edition, (Wiley-VCH, Weinheim 2007), quoted as EC and the respective page number. If necessary, further references to textbooks, review articles, and research papers are added.

A collection of experiments and their descriptions as provided following is a "work in progress". Further experiments related to new or less popular areas of electrochemistry will be added continuosly, for an update see http://www.tu-chemnitz.de/chemie/elchem/elpra. Further illustratins of experiments described in this book including short videos demonstrating visible changes of the system under investigation will be made available at the publishers website.

Symbols and descriptions in figures are used according to suggestions by IUPAC (Pure Appl. Chem. **37** (1974) 499). When compared with older textbooks this might occasionally result in minor confusion; the list of symbols, acronyms, and abbreviations will help (p. XV). Dimensions are separated by a slash (quantity calculus); square brackets are only used when necessary to avoid confusion.

Chemnitz, December 2008   Rudolf Holze

Foreword

Electrochemistry is everywhere in our daily life: It powers our cell phones, note-book computers and many other electronic devices, it provides the power to start our cars in the morning, it is undesirably present in corrosion, but also in metal winning and refining and the list goes on and on and...

Consequently electrochemistry is a prominent subject in science and technology. Being an experimental science, progress in electrochemistry strongly depends on the close interplay between theory and practice. This interplay should start in education and teaching as early as possible. The present book illustrates this suggestion nicely: It contains a wide collection of experiments covering almost all areas of experimental electrochemistry ranging from simple, basic experiments for classes in high school to elaborate ones for advanced students at universities. The book fills a gap that has existed for some time as indicated in the electrochemistry curriculum of the *International Society of Electrochemists*: There are numerous textbooks on electrochemistry, but a book on electrochemical experiments is completely missing. The broad selection of experiments, carefully and reproducibly described, offers numerous possibilities useful in different educational environments depending on local traditions in teaching, safety regulations and curricula; it favorably complements existing textbooks on this subject (without repeating their content in undue length). The obvious relationship between theory and experiment is consistently presented, and as an interesting feature of the book, the practical importance in our daily life is highlighted also in many experiments.

This book will hopefully be a helpful companion for everyone teaching and studying electrochemistry at all levels thus providing a deeper understanding of those numerous phenomena and processes that are associated with the science of electrochemistry.

Prof. R. Daniel Little
Department of Chemistry & Biochemistry
University of California
Santa Barbara
USA

Experimental Electrochemistry. A Laboratory Textbook. Rudolf Holze
Copyright © 2009 WILEY-VCH Verlag GmbH & Co. KGaA, Weinheim
ISBN: 978-3-527-31098-2

Symbols and Acronyms

A	Area
a	Activity
a_i	Debye length
C	Cell constant of a conductivity measurement cell
CV	Cyclic voltammogram
C_D	Double layer capacity
C_{diff}	differential double layer capacity
C_{int}	integral double layer capacity
c_p	isobar molar heat
c_V	isochoric molar heat
c	molar concentration
c_s	Surface concentration
c_0	Bulk concentration
D	Diffusion coefficient
d	Electrode distance
E	Electrode potential
E	electric field strength
ΔE_p	Difference of electrode peak potentials
E_0	Electrode potential at equilibrium with no flow of current, formal potential
E_{00}	Standard electrode potential
E_a	Energy of activation in a chemical reaction
E_F	Fermi energy, Fermi edge
$E_{Hg_2SO_4}$	see E_{MSE}
E_{MSE}	Electrode potential vs. a mercurous sulphate electrode, $c_{Hg_2SO_4}=0.1$ **M**
E_m	Measurement potential
E_{pzc}	Electrode potential of zero charge
$E_{p,ox}$	Electrode peak potential of oxidation reaction
$E_{p,red}$	Electrode peak potential of reduction reaction
E_{red}	Redox electrode potential
E_{ref}	Reference electrode potential
E_{RHE}	Electrode potential vs. relative hydrogen electrode

Experimental Electrochemistry. A Laboratory Textbook. Rudolf Holze
Copyright © 2009 WILEY-VCH Verlag GmbH & Co. KGaA, Weinheim
ISBN: 978-3-527-31098-2

E_{SCE}	Electrode potential vs. saturated calomel electrode SCE
e_0	Elementary charge
F	Force
F	Faraday constant
f	Measurement error, standard deviation; frequency, fugacity of a gas i ($f_i = \gamma_i p_i$)
ΔG	Gibbs energy (change); Gibbs energy of ion-solvent interaction
$\Delta H_{\text{Ion-LM}}$	Enthalpy of ion-solvent interaction
HOMO	Highest occupied molecular orbital
HRE	Hydrogen reference electrode
I	Ionic strength
I	Current (total current), also flow of species
I_a	Current transported by anions
I_c	Current transported by cations
I_C	Capacitive current
I_{ct}	Charge transfer current
I_{diff}	Diffusion-limited current (also: $I_{\text{lim,diff}}$)
I_D	Disc current at a ring-disc electrode
$I_{D,\,diff}$ D	iffusion limited disc current at a ring-disc electrode
I_p	Peak current
I_R	Ring current at a ring-disc electrode
$I_{R,\,diff}$	Diffusion limited ring current at a ring-disc electrode
I_{sc}	Short circuit current
j	Current density
j_{ct}	Charge transfer current density
j_{diff}	Diffusion limited current density (also: $j_{\text{lim,diff}}$)
j_{lim}	Limiting current density
j_R	Ring current density at a ring-disc electrode
j_D	Disc current at a disc or ring-disc electrode
K	Equilibrium constant
K_c	Concentration equilibrium constant, also: dissociation (equilibrium) constant
K_s	Dissociation (equilibrium) constant
k	Kohlrausch constant, rate constant
L	Conductance, electrical conductance; also: solubility product
LUMO	Lowest unoccupied molecular orbital
M	Molarity
M	Molar mass, atomic mass
m	Molality, flow rate of mercury at the dropping mercury electrode in $mg \cdot s^{-1}$
N_A	Avogadro's number (see also: N_L)
N_L	Loschmidt number (see also: N_A)
n	Number of mols
n	Electrode reaction valency

n_A	Number of mols of anions
n_C	Number of mols of cations
n^+	Stoichiometric coefficient of cations
n^-	Stoichiometric coefficient of anions
p. A.	pro analysi: pure for analysis, degree of purity of a substance
Q_{DL}	Electrical charge needed for double layer charging
q_e	Charge of an electron
q^-	Charge transported by anions
q^+	Charge transported by cations
R	Electrical resistance, gas constant
R_{ct}	Charge transfer resistance
R_f	Roughness factor
R_{sol}	Electrolyte solution resistance
RHE	Relative Hydrogen Electrode
r_i	Ionic radius
r_1	Disc radius of a ring-disc electrode
r_2	Inner ring radius of a ring-disc electrode
r_3	Outer ring radius of a ring-disc electrode
SOMO	Semioccupied molecular orbital
T	Absolute temperature
t	Transference number
t^+	Transference number of cations
t^-	Transference number of anions
t	Student's t-factor
U	Electrical voltage, difference of two electrode potentials
U_0	Electrical voltage at equilibrium ($I=0$), difference of two electrode potentials at equilibrium ($I=0$),
U_d	Decomposition voltage
u	Ionic mobility, $u=v/E$
V	Volume
v	Traveling velocity of ions; rate of mercury flow at a dropping mercury electrode
v	dE/dt, scan rate in cyclic voltammetry
v	Kinematic viscosity
x	Mole fraction
z	Ionic charge number

Greek symbols

α	Degree of dissociation, symmetry coefficient
χ	Surface potential
δ	Diffusion layer thickness
δ_N	Nernst diffusion layer thickness
$\varepsilon, \varepsilon_r$	Dielectric constant, relative dielectric constant
γ	Activity coefficient
φ	Volta potential
ϕ	Electrostatic potential
κ	Specific conductance
Λ_{eq}	Equivalent conductivity
Λ_0	Equivalent conductivity at infinite dilution
Λ_{mol}	Molar conductance
λ_{eq}^+	Equivalent ionic conductivity of cations
λ_{eq}^-	Equivalent ionic conductivity of anions
λ_{mol}^+	Molar ionic conductivity of cations
λ_{mol}^-	Molar ionic conductivity of anions
λ_0^-	Molar (equivalent) ionic conductivity of cations at infinite dilution
λ_0^+	Molar (equivalent) ionic conductivity of anions at infinite dilution
η	Overpotential
η_{ct}	Charge transfer overpotential
η	Dynamic viscosity
θ	Degree of coverage
ρ	Specific resistance
τ	Drop time of a dropping mercury electrode in s; transition time
ξ	Extent of reaction
ω	Angular velocity

1
Introduction – An Overview of Practical Electrochemistry

Students in natural sciences as well as professionals in numerous areas will meet electrochemical methods, concepts, and processes in many fields of science and technology. Accordingly, any a conceivable selection of possible experiments intended as an illustration of this width, the numerous possibilities of electrochemistry, and an introduction to the subject has to be similarly broad. According to the book's purpose and intention this will be achieved by the width of the selection of experiments, the scope of the practical (instrumental) requirements, and the necessary level of knowledge. Convenient use of the book and logical arrangement of the essentials of the theoretical introduction suggest a rational arrangement of experiments. As already proposed and executed elsewhere (R. Holze: Leitfaden der Elektrochemie, Teubner, Stuttgart 1998 and Elektrochemisches Praktikum, Teubner, Stuttgart 2000), electrochemistry in equilibrium, i.e. without flow of current and conversion of matter, is followed by electrochemistry with flow of current. In the first chapter, measurements of electrode potential and their application in, e.g., the determination of thermodynamic data are treated. The second chapter deals with all kinds of experiments where an electrical current crosses the electrochemical interface. Applications of electrochemical methods (both without and with flow of current) are handled in a chapter on electrochemical methods of analytical chemistry[1]. This chapter also contains experiments helpful in elucidating, e.g., mechanisms of electrode processes (a somewhat broader meaning of analytical) if the focus of the experiment is not on the experimental method itself, thus suggesting its inclusion in one of the preceding chapters. According to the growing impact of non-traditional, in particular spectroscopic, methods in electrochemistry a small section of experiments from this branch follows; unfortunately, the feasibility of these experiments depends crucially on the presence of mostly expensive instrument. Electrochemical methods of energy conversion and storage are of utmost practical importance, and numerous first personal interactions with electrochemistry

[1] The term analytical electrochemistry is neat and seemingly well-defined, but unfortunately in daily life its use is somewhat confusing: The term "analytical" is sometimes applied to qualify a certain branch of electrochemistry (in contrast, e.g., with synthetic electrochemistry); sometimes it means application of electrochemical methods in analytical chemistry – as intended here.

Experimental Electrochemistry. A Laboratory Textbook. Rudolf Holze
Copyright © 2009 WILEY-VCH Verlag GmbH & Co. KGaA, Weinheim
ISBN: 978-3-527-31098-2

deal with these devices. They combine applied aspects of equilibrium (i.e. thermodynamic) electrochemistry and electrochemical kinetics; consequently in this chapter only those experiments are collected where these aspects are not dominant. Electrochemical methods in industrial (synthetic) chemistry applied in the production of base chemicals like e.g. chlorine or sodium hydroxide and in synthetic procedures are subject of some experiments in the final chapter, the difficulties of the transfer of a large-scale industrial process into a simple laboratory experiment limit the selection.

The following descriptions of experiments are organized according to a general scheme. A brief statement of the experimenter's task and the aim of the experiment are followed by a condensed description of the theoretical foundations, essentially for understanding the experiment. This information cannot replace the respective parts of a textbook or the original reports in the primary literature. Besides references to primary sources, the respective sections of C. H. Hamann, A. Hamnett, and W. Vielstich: Electrochemistry (Wiley-VCH, Weinheim 2007) are quoted as EC and the respective page number: EC:xx. Some methods like polarography and cyclic voltammetry are employed in several experiments; nevertheless their fundamentals are described only once when the method is introduced first. No attempt is made in the descriptions to list all conceivable applications of the method used in this experiment. The tempting concept to arrange experiments according to difficulty or complexity of the experimental apparatus was discarded soon as being too personal and subjective. Instead, the reader and user of this book will easily select experiments according to his or her personal interests and intentions; the comparison of available and necessary equipment can subsequently been performed easily as well as the estimate of the required knowledge for successful execution.

The description of the execution of an experiment starts with a list of necessary instruments and chemicals. Possible alternative instrumentations are highlighted; the subsequent description is nevertheless limited to one experimental way only. The description contains – if necessary – a schematic circuit diagram of the setup and sketches of the construction of the apparatus or parts of it. The execution of the suggested measurements is briefly outlined. Potential pitfalls and unusual details are indicated. The way from the raw data to the desired results is sketched. Final questions including those pertaining to the practical execution help to confirm the newly acquired knowledge. Extensive calculation training examples are not included, these can be found in the textbook by J. O'M. Bockris and R. A. Fredlein. Typical results are displayed without cosmetic tidying up, this will encourage the user, it also demonstrates the level of skill needed to obtain satisfactory agreement between literature data (always quoted according to bibliographic standards from generally available textbooks for comparison) and one's own results.

Practical Hints

In most experiments aqueous solutions are used. If not stated otherwise ultra-pure water (sometimes called 18 MΩ-water because of the typical specific resistance value of this water) is used. It can be obtained by afterpurification of deionized water by various commercially available purification systems. As an alternative doubly distilled (bidestilled) water can be used. In some experiments simple deionized water can be used. Because especially in demanding conductivity [2] and potentiometric measurements traces of impurities present in deionized water may cause erroneous results, blind tests are required, in particular when water of less than ideal purity is used. In some cases not only the desired concentration of a necessary solution but also the amount of the selected chemicals needed for preparing the requested amount of solution are given for ease of preparation. When cells or other experimental setups with volumes different from the suggested setup are used these numbers must obviously be corrected. Purification of organic solvents has been thoroughly described by Mann (C. K. Mann: Nonaqueous Solvents for Electrochemical Use, Electroanalytical Chemistry **3** (A. J. Bard, ed.), Marcel Dekker, New York 1969, p. 57); further information on electrolyte solutions based on organic solvents has been collected by Gores and Barthel (H. J. Gores and J. M. G. Barthel, Pure & Appl. Chem. **67** (1995) 919).

Electrodes

As suggested by W. Nernst, the term electrode should always be applied to a specific combination of an electronically conducting material (e.g. a metal, graphite, a semiconductor) and an ionically conducting phase in contact with this material (e.g., an aqueous solution, a polymeric electrolyte, a molten salt). The need for this use will be neatly illustrated in experiment 3.13 with lead being in contact with various electrolyte solutions; quite obviously the term lead electrode becomes ambiguous. In daily life the term electrode mostly refers to the electronically conducting component only. This well-established usage will not be completely suppressed in this book; nevertheless the possible confusion will be addressed repeatedly.

---

2) Meaning and use of the terms conductance and conductivity and their respective reciprocal relatives resistance and resistivity are slightly confusing and frequently inconsistent. According to standard reference works conductance applies to the conducting power of matter without requiring specific dimensions of the sample whereas conductivity refers to the conductance of a sample with unit dimensions, i.e. it refers to specific conductance. The frequently employed term "specific conductivity" might thus be popular, but it remains a pleonasm or tautology. In this book the standard terms conductivity meter or conductivity measurements are employed – although in most cases only conductances are measured.

In some experiments electrodes of special shape and construction prepared from selected materials are needed; details are provided in the descriptions of the experiments. In many experiments electrodes of a fairly general type and construction will be used. Because they can be prepared easily in a glassblower's shop or even without any expert help, some suggestions are given below.

Frequently, metal sheets (of noble metals like platinum or gold) are used as working and counter electrodes. These electrodes can be manufactured easily by spot welding a metal wire to a piece of sheet metal (about 0.1 to 0.2 mm thick). After extending the metal wire with a piece of copper wire connected with hard solder (i.e. silver solder; soft solder is not recommended because it will most likely melt in the subsequent glassblowing operation; in addition soft solder forms alloys with gold making a reliable connection impossible) the noble metal wire is sealed into a glass tube. Glass with low melting point is preferred because the low viscosity of the molten glass obtained even at moderate temperatures provides a tight glass-metal seal. With platinum, borosilicate glass of higher melting point can be used. Particularly useful even for silver wires are lead dioxide-based glasses. The glassblower must avoid reducing conditions when operating his blowtorch. Unfortunately these glasses are hard to get. When a spot welding machine is not available simple metal wire spirals can be used instead of sheet electrodes. Instead, with a glass-metal seal the wire can also be fixed with epoxy glue; unfortunately this connection is mechanically and chemically less stable, may not be exposed to some aggressive cleaning solutions, and in addition may release traces of contaminants into the electrolyte solution.

In the case of a very simple preparation of a microelectrode this epoxy glue is essential as shown below (Fig. 1.1). A glass tube is heated until its end almost collapses and a very narrow opening remains. A carbon fiber is fed through the opening. The opening is closed with epoxy. By pulling and pushing the fiber gently it is coated with epoxy and completely surrounded with the glue. After curing the epoxy an electric connection to a copper wire is provided on the inside with conductive silver or graphite cement.

As reference electrode mostly hydrogen or metal ion electrodes are used. Saturated calomel and silver chloride electrodes (EC:99) are particularly popular. Figure 1.2 shows typical constructions.

Various aqueous as well as nonaqueous electrolyte solutions can be used; in all cases the high chloride concentration may result in contamination of the electrolyte solution in the electrochemical cell because of halide diffusion

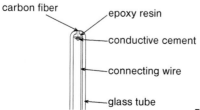

carbon fiber — epoxy resin

conductive cement

connecting wire

glass tube

Fig. 1.1 Cross section of a simple microelectrode.

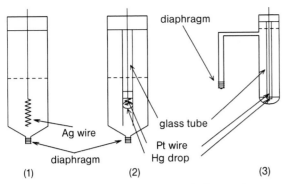

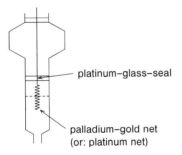

Fig. 1.2 Cross sections of various types of reference electrodes: (1) Silver/silver chloride electrodes; (2, 3) Calomel or mercurous sulphate electrode, dotted lines indicate solution filling level.

platinum–glass–seal

palladium–gold net
(or: platinum net)

Fig. 1.3 Hydrogen electrode according to Will.

through the separating diaphragms etc. In alkaline solution disproportionation of calomel may occur.

A particularly simple design of a hydrogen electrode[3] as suggested by Will (F. G. Will and H. J. Hess, J. Electrochem. Soc. **120** (1973) 1; F. G. Will and H. J. Hess, J. Electrochem. Soc. **133** (1986) 454) is shown below (Fig. 1.3).

Platinum wire net (better: a palladium or palladium/gold wire net) is spot-welded to a platinum wire. This assembly is fixed inside the glass tube drawn into a capillary with a glass-metal seal. The active surface area of the net is enlarged by electrolytic coating with platinum black. Using the electrolyte solution of the planned experiment, an auxiliary electrode (e.g. platinum wire) as anode

3) Sometimes hydrogen electrodes are called reversible or relative hydrogen electrodes RHE. The former term is obviously redundant. The term reversible implies the presence of a reversible reaction: An electrode reaction proceeds on the same reaction pathway in both directions at significant rates; without this no stable electrode potential would be established. Alternatively the term might imply reversible in a thermodynamic sense – being at equilibrium. Because a reference electrode is always used without any flow of current it is in equilibrium. The term relative refers to the fact that a nonstandard proton concentration may be present, thus the electrode is not a standard electrode.

and a DC power supply (a few volts will be enough), the metal net inside the capillary is charged with hydrogen by cathodic reduction of protons until a hydrogen gas bubble is formed. The bubble may stay in place for several weeks when the glass-metal seal is sufficiently tight; accordingly the electrode potential remains stable. In the case of organic compounds dissolved in the electrolyte solution, which may react at the platinum surface, potential shifts may occur as a consequence of electrode poisoning; in the case of neutral electrolyte solution the potential is also less stable because the exchange current density of the hydrogen electrode reaction is pH-dependent (EC:344). The activity of protons in the solution filled into the hydrogen electrode will be unity only in rare cases. Consequently this electrode may not be called a standard hydrogen electrode SHE (the term normal hydrogen electrode NHE should be avoided altogether because the term normal may be taken as designation of a certain concentration); because the hydrogen electrode potential is related to the proton activity this hydrogen electrode is sometimes called relative hydrogen electrode RHE.

With nonaqueous electrolyte solutions the use of reference electrodes containing aqueous electrolyte solutions is still possible with salt bridges (see below) providing the ionically conducting connection. The diffusion potentials created at the phase boundaries between the various solutions may cause considerable experimental errors if not corrected properly. In addition the slow diffusion of solution components (including of course water) into the nonaqueous electrolyte solution may cause undesirable chemical reactions or other experimental artifacts. Thus electrolyte solutions employing nonaqueous solvents are an attractive option. Unfortunately electrodes of the second kind (e.g., silver/silver chloride, for details see above) are prone to disproportionation and finally decomposition resulting in reference electrode potential drift. In addition the potential of these electrodes depends on the employed solvent and thus comparison between experimental results obtained with different solvents may be unreliable. Redox systems with a formal potential E_0 defined as the midpoint potential between their respective oxidation and reduction peak potentials as observed in a cyclic voltammogram $(E_0 = E_{p,\,red} + (E_{p,\,ox} - E_{p,\,red})/2$, for details see experiment 3.18) have been employed as point of reference repeatedly provided that the value of E_0 does not depend on the solvent. Ferrocene (see experiment 3.18) has been suggested as a candidate (R. R. Gagné, C. A. Koval, and G. C. Lisensky, Inorg. Chem. **19** (1980) 2854) because the iron ion in the center between the two cyclopentadienyl ligands seems to be shielded fairly well from the solvent. Unfortunately it has turned out that upon oxidation of ferrocene most of the charge is removed from the ligand; the actual charge on the iron is changed only by about $1/10^{th}$ of an electron charge. As a substitute decamethyl ferrocene or decaphenyl ferrocene have been suggested; for a review see I. Noviandri, K. N. Brown, D. S. Fleming, P. T. Gulyas, P. A. Lay, A. F. Masters, and L. Phillips, J. Phys. Chem. B **103** (1999) 6713. In an actual experiment a reference electrode of the second kind filled with a nonaqueous electrolyte solution may be used; at the end of the experiment some ferrocene is added, a cyclic voltammogram is recorded, and all electrode potentials are converted to this reference (e.g. the ferrocene) scale. Only in this way can results obtained with different solvents at different places be made compatible and comparable and comparisons be made possible.

Although a large selection of electronic reference voltage sources with a precision sufficient for calibration purposes is available (for a suggested circuit see the appendix) electrochemical reference cells are still in use. The only infrequently encountered Clark cell ($Zn|ZnSO_{4,(aq., sat.)}|ZnSO_{4, sol.}+Hg_2SO_{4, sol.}|Hg$) has been replaced by the Weston cell ($Cd|CdSO_{4,(aq., sat.)}|CdSO_{4, sol.}+Hg_2SO_{4, sol.}|Hg$). The cell reaction of the latter has significantly lower entropy of reaction resulting in a smaller temperature coefficient of the cell voltage.

Measuring Instruments [4]

Aside from special instruments typical of certain experimental methods (e.g., polarographs for polarography), simple electronic instruments are frequently needed in particular for the measurement of voltage and current. Standard analog or digital multimeters are sufficient in most cases. During measurements of electrode potentials performed as the measurement of the voltage between a reference electrode and the working electrode under study the flow of electric current must be avoided. Ideally this has been done using a compensation circuit. This procedure is cumbersome and today of only limited practical importance. Voltmeters with very high input resistance ($R_i > 10^{12}\,\Omega$) are sufficiently close to this ideal method. During selection of instruments for precision measurements this input resistance merits particular attention. Budget-priced multimeters are frequently built with a voltage divider circuit for range selection at the input resulting in fairly low input resistance values; these should be used with appropriate respect. With commercially available digital voltmeter modules (no input voltage divider added) with a 2-V range a powerful instrument for potential measurements can be built at low cost. Measurements of current-potential curves require current meters with a very small measuring (shunt) resistor; ideally its value should be zero. This can be realized with simple circuits based on operational amplifiers (current followers) which are particularly helpful during measurements with electrochemical cells providing only small output voltages (fuel cells, batteries). Because of lower price and apparently higher precision, digital instruments are frequently preferred. During experiments wherein a voltage has to be adjusted to a certain value an analog instrument may be better because trends in the measured signal can be discerned more clearly. Digital instruments with an additional bargraph display may be a compromise; it may be difficult to become accustomed to the flickering bargraph display.

4) A complete setup including a potentiostat built as plugin for a standard desktop PC, a cell, several electrodes, and software suitable for running several of the experiments described below including a workbook is offered by Sycopel Scientific Instruments, 15 Sedling Road, Washington NE38 9BZ, Great Britain.

Electrochemical Cells

As well as specific cells designed just for a certain experiment as described in the text related to it, some generally used types of electrochemical cells have come into use. A simple beaker will be sufficient only in a few cases because operation with inert gas atmosphere above the electrolyte solution and sufficient purging of the solution itself are impossible with this arrangement. In addition reliable and reproducible mounting of electrodes can be done only with additional holders etc. For cyclic voltammetry a cell depicted below in cross section, commonly called H-cell (because of its shape) is frequently used (Fig. 1.4).

For electrochemical impedance measurements and further experiments with AC-modulation of the electrode potential spherical working electrodes or circular disc-shaped electrodes embedded in an inert material mounted in the center of a symmetric cell vessel are advantageous. Additional ground glass feed-throughs hold glass tubes for gas purge of the electrolyte solution and the counter and reference electrode. Reduced exchange of electrolyte solution between the interior of the latter tubes and the main compartment of the cell can be achieved by closing the tubes with porous glass or ceramic frits. This type of cell can also be employed in studies with a rotating disc electrode as shown below (Fig. 1.5).

In precision studies[5] and in experiments, where mixing of electrolyte solutions must be prevented, "salt bridges" (Fig. 1.6) are needed. They provide electrolytic connection between cell compartments without bringing the electrolyte solutions in these compartments into direct contact. In a very simple design a piece of plastic tube filled with a suitable electrolyte solution and closed with cotton wool plugs or plugs made of filter paper may be sufficient. The electrolyte could be 1 **M** KNO_3, both ions having similar mobility (P. W. Atkins, J. de Paula, Physical Chemistry, 8th ed., Oxford University Press, Oxford 2006, p. 1019) which results in an approximate compensation of diffusion potentials (P. W. Atkins, J. de Paula, Physical Chemistry, 8th ed., Oxford University Press, Oxford 2006, p. 216; EC:112;146) generated at the interfaces the cell solutions.

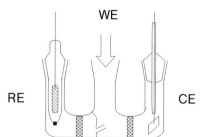

WE

RE   **CE**

Fig. 1.4 Cross section of an H-cell for electrochemical experiments, WE: working electrode, RE: reference electrode, CE: counter electrode.

---

5) For example measurements of standard electrode potentials or cell voltages wherein diffusion potentials must be either avoided or precisely known.

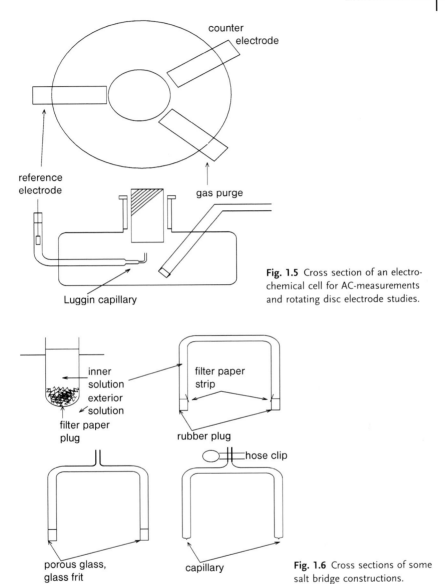

Fig. 1.5 Cross section of an electro-chemical cell for AC-measurements and rotating disc electrode studies.

Fig. 1.6 Cross sections of some salt bridge constructions.

In addition both ions show only minor interference with most electrochemical processes and cell reactions. More reliable and stable are ground glass diaphragms or porous glass plugs (made of "VYCOR®"[6]). Some typical designs are shown above (Fig. 1.6).

---

6) These plugs must be kept wet, as drying might result in destruction.

Data Recording

Extensive data recording may not be necessary with most experiments. Only in studies of processes where data have to be recorded at a high rate (e.g., in cyclic voltammetry or spectroscopy) are recording devices needed. In addition to oscilloscopes (which provide printed output in most cases only with added hard- and software at a considerable price), analog recorders (X-Y-recorders or X-t-recorders) have been used. The omnipresence of cheap and powerful computers and the steep rise in the price of analog recorders suggest the use of computerized data recording setups. Easily programmable and cheap ADDA-interface[7] cards enable the generation of input signals for electrochemical experiments (via the DA-capability) and the recording (via the AD-capability) making the computer a replacement for both recorder and function generator. Subsequent data storage and treatment would preferably be by the use of a computerized workplace instead of cumbersome paper shuffling. Some drawbacks should be kept in mind nevertheless.

Commercial software for electrochemical experiments is powerful and expensive in most cases (for an exception see footnote on p. 8), so that its purchase just for a laboratory course may be inappropriate. In addition the use of analog function generator and X-Y-recorder has some advantage when adjusting experimental parameters like, e.g., potential limits, scan rates etc. because the effect will become visible immediately. With computerized setups this is impossible in most cases; thus the direct cause-effect relationship important in education and hands-on teaching gets lost. Interface cards need careful hard- or software calibration, whichs tends to be forgotten frequently. Because calibration needs external voltage sources providing unusual voltages (not the values of standard cells described above) a simple circuit based on a high-precision voltage reference circuit is provided in the appendix. This circuit needs only a single initial calibration versus a high-precision reference. When selecting measuring ranges on potentiostats and other devices the input voltage range of the DA-interface card must be kept in mind. The range is frequently 2 V or 5 V. In potential measurements (actually the voltage between the working and the reference electrode is measured by the potentiostat) this range is a very good match. In the case of current measurements, frequently only small currents are observed. Unsuitably small shunt resistors provide only very small voltages resulting in a poor signal-to-noise ratio on the current display. Thus a shunt resistor as a high as possible should be selected; in modern potentiostats the built-in current follower permits this good match.

---

7) ADDA: analog-digital/digital-analog converter

2
Electrochemistry in Equilibrium

In electrochemistry an organizing criterion frequently employed is the flow of current: Experiments or – more generally – phenomena without current in the system under investigation naturally, at equilibrium, may be related both to the interior (bulk) of the electrolyte solution and to the electrochemical interface between electron conducting-material (e.g., metal, conveniently also incorrectly called electrode) and ionically conducting electrolyte solution. Both possibilities exist also in dynamic systems. In this case the flow of electronic current in the wires, metals etc. is coupled to the flow of ions in the ionically conducting phase (electrolyte solution). Both processes and fluxes are coupled at the electrochemical interface.

In this chapter, experiments without flow of current are described. Fundamental facts and relationships of electrochemical and general thermodynamics, mixed phase and non-ideal thermodynamics, and relationships between electrochemical and thermodynamic data are discussed.

Experiment 2.1: The Electrochemical Series

Task
A standard hydrogen electrode is prepared and is used as a reference in the determination of the standard potentials of nickel, copper, and zinc electrodes. The influence of the metal ion concentration as suggested by the Nernst equation is examined. The temperature dependence of the voltage of a copper-silver cell is measured and used for the calculation of the entropy of reaction.

Fundamentals
Between the chemical elements and their compounds substantial differences exist in their tendency to be reduced (with associated uptake of electrons) or oxidized (with respective removal of electrons). In electrochemistry the comparison between these properties for two elements can be performed most easily by measuring a cell voltage. Comparable conditions, in this case standard ones, are established by using the elements, in particular metals (which are studied here)

Experimental Electrochemistry. A Laboratory Textbook. Rudolf Holze
Copyright © 2009 WILEY-VCH Verlag GmbH & Co. KGaA, Weinheim
ISBN: 978-3-527-31098-2

as electrodes in solutions containing their ions at unit activity. Because of the nonideal behavior of ions in solution, this activity $a=1$ **M** is mostly reached only with concentrations considerably larger than unity ($c>1$ **M**). Between the solutions containing the hydrogen reference electrode and the metal electrode a salt bridge is employed. Cell voltage between the metal wire connector of the hydrogen electrode and the metal of the second electrode is measured with a high input impedance voltmeter. This voltage can be compared with the voltage calculated based on the cell reaction and the respective electrode reaction deduced by splitting the cell reaction. The metal identified as the plus terminal (cathode) is called nobler than the other metal, which constitutes the minus pole (anode). Taking the less noble zinc and the more noble copper as examples this can be verified. Taking solutions with the respective ions at standard activity the zinc wire is the minus pole, and the copper wire the plus pole. This cell is known as the Daniell element (cell). Reactions are:

Cathode (Reduction): $Cu^{2+} + 2e^- \rightarrow Cu$ (2.1)

Anode (Oxidation): $Zn \rightarrow Zn^{2+} + 2e^-$ (2.2)

Cell reaction: $Cu^{2+} + Zn \rightarrow Zn^{2+} + Cu$ (2.3)

Comparative measurements permit the establishment of a list showing metals (and more generally speaking elements and compounds) rated according to their reductive or oxidative capability. This list is called the electrochemical series. These measurements are also possible with gaseous reactants. Thus a hydrogen electrode is established by bubbling hydrogen gas around an inert metal electrode (e.g., a platinum sheet) in contact with an aqueous electrolyte solution of well-defined pH-value. With proton activity $a=1$ and hydrogen pressure $p=p_0=1$ atm (= 101 325 Pa) a standard hydrogen electrode is obtained. Measurements with this hydrogen electrode as one electrode (or half-cell) and another test electrode yield cell voltages equivalent to the electrode potential of the test electrode because by definition the electrode potential of the standard hydrogen electrode is $E_{SHE}=0$ V. This value is equivalent to the number given in the electrochemical series.

With electrolyte solutions of other ionic activities and gas pressures (in case of gas electrodes where the gas is involved in establishing the electrode potential), different, non-standard values of the electrode potential are obtained. The relationship between activities, pressures and electrode potential is given by the Nernst equation.

The cell voltage is related to the Gibbs energy of the cell reaction according to

$$\Delta G_R = -z \cdot F \cdot U_0$$ (2.4)

Using the partial derivative of the Gibbs equation assuming a constant (i.e. temperature-independent) value of the reaction enthalpy ΔH_R in the studied range of temperatures

$$(\partial \Delta G_R / \partial T)_p = (\partial \Delta H_R / \partial T)_p - (\partial T \Delta S_R / \partial T)_p \qquad (2.5)$$

yields

$$(\partial \Delta G_R / \partial T)_p = -\Delta S_R \qquad (2.6)$$

The reaction entropy can now be calculated from measurements of the temperature coefficient of the cell voltage:

$$(\partial U_0 / \partial T)_p \cdot z \cdot F = \Delta S_R \qquad (2.7)$$

Execution
Chemicals and instruments
Aqueous solution of $CuSO_4$, 1 **M**
Aqueous solution of $ZnSO_4$, 1 **M**
Aqueous solution of $NiSO_4$, 1 **M**
Aqueous solution of $AgNO_3$, 1 **M**
Aqueous solution of HCl, 1.25 **M** (the proton activity is approx. 1)
Salt bridge filled with 1 **M** KNO_3 [1]
Silver, copper, nickel, platinum, and zinc electrodes (wires, sheets)
Hydrogen gas
High input impedance voltmeter
Thermostat

Setup
The metal salt solutions are filled into beakers; the metal electrodes are cleaned with abrasive paper, rinsed with water and immersed in the respective solutions. Connection between two beakers is made with the salt bridge; when the beakers are exchanged the tips of the salt bridge are careful rinsed with water to avoid contamination.

Procedure
The voltage between the two metal terminals is measured with the various possible combinations of electrodes. In addition, the cell voltage of the metal electrodes vs. the hydrogen electrode is measured.

The zinc sulfate solution is diluted to 0.1 **M** and 0.01 **M**; the measurements vs. the hydrogen electrode are repeated.

---

[1] Filling the salt bridge with a solution of KCl is not recommended because chloride ions adsorb specifically on most metals and may cause corrosion.

Using the cell Ag/AgNO$_3$ solution/salt bridge/CuSO$_4$ solution/copper the cell voltage as a function of temperature in the range 20 °C $< T <$ 80 °C is measured[2].

Evaluation

The obtained values are listed as in the electrochemical series; they are compared to literature data. Values obtained with zinc sulfate solutions of various concentrations are examined using the Nernst equation.

Measurement of the cell voltage of the copper-silver cell in an easily accessible range of temperatures results in data as displayed for a typical experiment below (Fig. 2.1). Conceivable causes of deviations between obtained numbers and expected ones are differences in temperature between the interior of the cell and the thermostat bath. The calculated value[3] for room temperature RT is $U_0 = 0.469$ V; the experimentally observed one is $U_0 = 0.454$ V. To identify sources of this difference, the potentials of both electrodes can be measured versus a reference electrode. With a saturated calomel electrode the results are: $E_{Cu\ vs.\ SCE} = 0.322$ V and $E_{Ag\ vs.\ SCE} = 0.775$ V. Obviously the deviation is caused by a non-ideal behavior of the copper electrode.

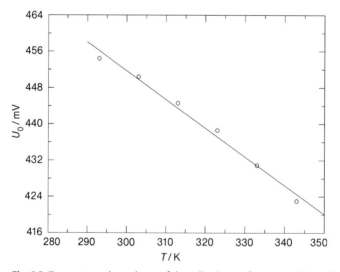

Fig. 2.1 Temperature dependence of the cell voltage of a copper-silver cell.

2) Measurements of the temperature coefficient of the Daniell cell seem to be a more attractive way to obtain reaction entropy because thermodynamic data for this cell reaction are well known. The very small reaction entropy of this reaction has already induced earlier researchers to assume that the Daniell cell provides a way for complete conversion of the reaction enthalpy into useful work. The copper electrode tends to form poorly defined surface oxide layers especially when exposed to corrosive environments. The amphoteric nature of the zinc electrode adds further uncertainty, making this cell a less attractive candidate.

3) Activity coefficients are $\gamma_{CuSO_4} = 0.047$ and $\gamma_{AgNO_3} = 0.4$.

Using the temperature coefficient calculated from the plot $\partial U_0/\partial T = -0.63$ mV K^{-1} a reaction entropy $\Delta S = -121$ J·K^{-1}·mol^{-1} can be calculated. The value calculated from thermodynamic data (P. W. Atkins, J. de Paula, Physical Chemistry, 8th ed., Oxford University Press, Oxford 2006, p. 995) is $\Delta S = -145$ J·K^{-1}·mol^{-1}.

Questions
- Can an electrode be established with an insulator instead of a metal?
- Is the answer valid also for semiconductors (e.g. silicon)?

Experiment 2.2: Standard Electrode Potentials and the Mean Activity Coefficient

Tasks
- By measuring the cell voltage of the galvanic cell Ag/AgCl/HCl/H$_2$/Pt the following are determined:
  a: standard electrode potential of the silver chloride electrode
  b: the mean activity coefficient of HCl in aqueous solutions.
- The electrode potential of the silver ion electrode and the Fe^{2+}/Fe^{3+}-redox electrode are measured as a function of concentration.

Fundamentals
The electrode potential is by definition the voltage measured between the electrode under investigation and a standard hydrogen electrode. If both electrodes are in standard state (i.e. unit activities of all species controlling the electrode potential and room temperature) the electrode potential is a standard potential. Experimental determination of standard potentials yields activities calculated with the Nernst equation; this in turn provides access to several thermodynamic data (e.g. equilibrium constants, activity constants).

In this experiment particular attention must be paid to measurement of voltages at zero current and without diffusion potentials. High input impedance voltmeters are necessary; in addition, calibration vs. a Weston cell is recommended.

Diffusion potentials can be avoided completely only when both the hydrogen electrode and the electrode under study are immersed in the same electrolyte solution (cell without transfer; EC:108). When the electrolyte solutions in the electrodes differ, salt bridges should be used filled with electrolyte solutions of salts with equal mobility of anions and cations like, e.g., KCl or KNO$_3$.

The use of a hydrogen electrode is somewhat inconvenient because permanent purging of the platinum electrode with a stream of hydrogen gas is necessary. Thus, other electrodes with a stable, reproducible and well-defined electrode potential are frequently used instead. Particularly popular are calomel and silver chloride electrodes. Their electrode potential depends on the concentration of KCl employed as electrolyte (electrodes of the second kind) and the respective values are tabulated (R. Holze: Landolt-Börnstein: Numerical Data and

Functional Relationships in Science and Technology, New Series, Group IV: Physical Chemistry, Volume 9 A: Electrochemistry, Subvolume A: Electrochemical Thermodynamics and Kinetics, W. Martienssen, M.D. Lechner Eds, Springer-Verlag, Berlin 2007).

The relationship between electrode potential (always in equilibrium in this chapter) and activities of the species involved in the electrode reaction is given by the Nernst equation:

$$E = E_0 + ((R \cdot T)/(n \cdot F)) \cdot \ln \prod_i a_i^{\nu_i} \tag{2.8}$$

The product of activities $\prod_i a_i^{\nu_i}$ equals the equilibrium constant of the electrode reaction. Activities of solid phases are unity; the same applies to gases at standard pressure. Accordingly the following relationships can be derived for the electrodes studied here:

a) Ag/Ag$^+$-Electrode

$$Ag \leftrightarrows Ag^+ + e^- \tag{2.9}\,^{4)}$$

$$E_0(Ag/Ag^+) = E_{00}(Ag/Ag^+) + ((R \cdot T)/F) \cdot \ln a_{Ag^+} \tag{2.10}$$

b) Fe^{2+}/Fe^{3+}-Electrode

$$Fe^{2+} \leftrightarrows F^{3+} + e^- \tag{2.11}$$

$$E_0(Fe^{2+}/Fe^{3+}) = E_{00}(Fe^{2+}/Fe^{3+}) + (R \cdot T/F) \cdot \ln (a_{Fe^{3+}}/a_{Fe^{2+}}) \tag{2.12}$$

c) H$_2$-Electrode

$$H_2 \leftrightarrows 2H^+ + 2e^- \tag{2.13}$$

$$E_0(H_2) = E_{00}(H_2) + ((R \cdot T)/(2 \cdot F)) \cdot \ln (a_{H^+}^2/p_{H_2}) \tag{2.14}$$

$$= E_{00}(H_2) + ((R \cdot T)/(F)) \cdot \ln(a_{H^+}/p_{H_2}^{1/2}) \tag{2.15}$$

or at $p_{H_2} = p_0$ and $E_{00}(H_2) = 0$ V

$$E_0(H_2) = ((R \cdot T)/F) \cdot \ln a_{H^+} \tag{2.16}$$

d) Ag/AgCl-Electrode

$$Ag + Cl^- \leftrightarrows AgCl + e^- \tag{2.17}$$

$$E_0(Ag/AgCl) = E_{00}(Ag/AgCl) + ((R \cdot T)/F) \cdot \ln (a_{AgCl}/a_{Ag} \cdot a_{Cl^-}) \tag{2.18}$$

---

4) For better visibility, characters used to identify an electrode are placed on the line instead of as subscripts.

$$E_0(Ag/AgCl) = E_{00}(Ag/AgCl) - ((R \cdot T)/F) \cdot \ln a_{Cl^-} \qquad (2.19)$$

Standard electrode potentials are needed for calculations of activity coefficients based on electrode potential measurements. The following procedure is applied:

1. Measurement of equilibrium cell voltage at various concentrations of ionic species
2. Extrapolation of data to $c = 0$ **M** ($\gamma = 1$) based on Debye-Hückel theory.

As an example the cell

$$Ag/AgCl/HCl/H_2/Pt \qquad (2.20)$$

is considered with the cell reaction

$$AgCl + 1/2\,H_2 \leftrightarrows Ag + H^+ + Cl^- \qquad (2.21)$$

The cell voltage $U_0{}^{5)}$ is given by

$$U_0 = E_0(Ag/AgCl) - E_0(H_2) \qquad (2.22)$$

$$= E_{00}(Ag/AgCl) - ((R \cdot T)/F) \cdot \ln a_{Cl^-} - ((R \cdot T)/F) \cdot \ln a_{H^+} \qquad (2.23)$$

$$U_0 - E_{00}(Ag/AgCl) = -((R \cdot T)/F) \cdot (\ln a_{Cl^-} + \ln a_{H^+}) \qquad (2.24)$$

With $a_{Cl^-}^2 = a_{H^+} \cdot a_{Cl^-}$

$$U_0 - E_{00}(Ag/AgCl) = -((R \cdot T)/F) \cdot \ln a_{HCl}^2 \qquad (2.25)$$

$$U_0 - E_{00}(Ag/AgCl) = -((2 \cdot R \cdot T)/F) \cdot \ln a_{HCl} \qquad (2.26)$$

and $a_{HCl} = c_{HCl} \cdot \gamma_{HCl}$

$$U_0 - E_{00}(Ag/AgCl) = -((2 \cdot R \cdot T)/F) \cdot \ln(c_{HCl} \cdot \gamma_{HCl}) \qquad (2.27)$$

$$U_0 - E_{00}(Ag/AgCl) = -((2 \cdot R \cdot T)/F) \cdot \ln c_{HCl}$$

$$-((2 \cdot R \cdot T)/F) \cdot \ln \gamma_{HCl} \qquad (2.28)$$

and according to Debye-Hückel theory $\ln \gamma_{\pm} = -\,0.037 c^{1/2}$

$$U_0 - E_{00}(Ag/AgCl) = -((2 \cdot R \cdot T)/F) \cdot \ln c_{HCl} + ((2 \cdot R \cdot T)/F) \cdot 0.037\ c^{1/2} \qquad (2.29)$$

$$U_0 + ((2 \cdot R \cdot T)/F) \cdot \ln c_{HCl} = E_{00}(Ag/AgCl) + ((2 \cdot R \cdot T)/F) \cdot 0.037\ c^{1/2} \qquad (2.30)$$

Plotting $(U_0 + ((2 \cdot R \cdot T)/F) \cdot \ln c_{HCl})$ versus $c^{1/2}$ yields the standard electrode potential $E_{00}(Ag/AgCl)$ as the y-axis intersection.

5) Contact potential differences caused by differences in work functions of the employed metals are neglected here.

Execution

Chemicals and instruments

Dilute nitric acid (1:1)

Aqueous solution of $AgNO_3$, 0.1 **M**

Aqueous solution of KNO_3, 0.1 **M**

Aqueous solution of $FeCl_3$, 0.01 **M**, in 0.1 **M** HCl

Aqueous solution of $FeSO_4$, 0.1 **M**, in 0.1 **M** H_2SO_4

Aqueous solution of $FeSO_4$, 0.01 **M**, in 0.1 **M** H_2SO_4

Aqueous solution of HCl, 3 **M**, in automatic burettes

Galvanic cell with H_2- and Ag/AgCl-electrode

Hydrogen supply (tank, pressure reducer, needle valve)

Nitrogen supply (tank, pressure reducer, needle valve)

High input impedance voltmeter

Galvanometer (sensitive ammeter)

Weston cell

Silver electrode

Platinum electrode

Calomel electrode

10 measurement flasks 100 ml

1. Standard Potential and Mean Activity Coefficient

Setup

The setup of the galvanic cell is shown below (Fig. 2.2):

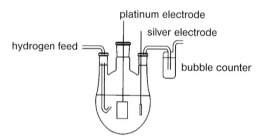

Fig. 2.2 The galvanic cell for determination of a standard potential.

Procedure

- By diluting the stock solution of 3 **M** HCl the following solutions (10 ml each) are prepared: 2, 1, 0.5, 0.1, 0.05, 0.01, 0.005, 0.001, and 0.0005 **M**.
- The galvanic cell is filled with one of the solutions starting with the most dilute one; hydrogen flow is adjusted to about 2 bubbles per second.
- The voltmeter is calibrated (if necessary) with the Weston cell.

- Because of the logarithmic relationship between activity and electrode potential determination of the equilibrium potential must be performed with great care. Cell voltages must be constant within 0.1 mV before recording.

Evaluation

The standard potential of the Ag/AgCl-electrode is determined graphically according to eq. (2.30). Figure 2.3 shows a typical result:

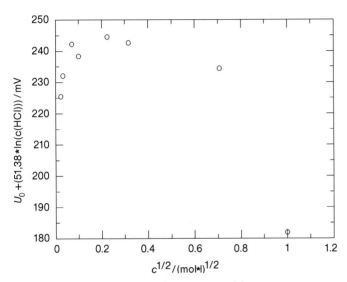

Fig. 2.3 Plot used for the graphic determination of the standard electrode potential of the Ag/AgCl-electrode.

Extrapolation using the cell voltages measured at small concentrations (at higher concentrations deviations are too large) yields a value of E_{00} =0.225 V; the literature value is E_{00} =0.222 V.

Using this result, activities a and activity coefficients γ of HCl can be calculated according to

$$U_0 = E_{00}(Ag/AgCl) - ((2 \cdot R \cdot T)/F) \cdot \ln a_{HCl} \qquad (2.31)$$

With the data displayed values of γ ranging from 0.904 at the lowest and 5 at the highest concentration are calculated.

2. Determination of Electrode Potentials

Setup

The silver electrode is formed by a silver wire dipping into an aqueous solution of $AgNO_3$. The saturated calomel electrode serving as the second (and reference) electrode is connected to the silver electrode via a salt bridge filled with an aqueous solution of 0.1 M KNO_3.[6)]

A platinum electrode dipping into the solution containing both Fe^{2+} and Fe^{3+} ions establishes a redox electrode. The saturated calomel reference electrode can be dipped directly into this solution.

The reference electrode is always connected to the "low" or "mass" or "ground" input of the voltmeter in order to get voltages and potentials with proper sign.

Execution

Ag/Ag$^+$-electrode
Diluted aqueous solutions (25 ml each) of $AgNO_3$ with concentrations of 0.05, 0.02, 0.01, 0.005, 0.002, and 0.001 M are prepared by diluting the stock solution. The silver wire is cleaned with dilute nitric acid, carefully rinsed, and dipped into the silver nitrate solutions. Voltages are measured with a high input impedance voltmeter versus the saturated calomel electrode connected to the silver electrode with the salt bridge, starting with the most diluted solution.

Fe^{2+}/Fe^{3+}-electrode
25 ml of a 0.01 M solution of $FeCl_3$ are placed in the cell vessel and purged with nitrogen. Subsequently 0.5, 1, 5, 10, and 20 ml of the 0.01 M aqueous solution of $FeSO_4$ are added; after every addition the solution is mixed with a brief nitrogen purge and the voltage versus the saturated calomel reference electrode is measured after switching off the purge. In order to obtain low ratios of c_{ox}/c_{red} the procedure is repeated starting with 25 ml of a 0.1 M solution of $FeCl_3$ and subsequent additions of 2.5, 5, 10, and 25 ml of a 0.1 M aqueous solution of $FeSO_4$.

Evaluation

From the measured cell voltages the electrode potentials are calculated ad plotted vs. lg c_{Ag^+} or lg$(c_{Fe^{3+}}/c_{Fe^{2+}})$. The slope of the curves is determined d extrapolation to lg $c=0$ is done to determine the respective standard electrode potentials. Taking into account the potential of the saturated calomel electrode, the standard potential of the silver electrode is determined: $E_{00}=0.81$ V. The literature value is $E_{00}=0.799$ V. A typical plot of the measured cell voltage is displayed in Fig. 2.4.

Questions
- Explain the term "mean activity coefficient"
- Describe the calculation of mean activity coefficients according to the Debye-Hückel theory.
- Explain the term "cell without transference"

---

6) Contamination of the silver nitrate solution with chloride ions
must be avoided; the porous plug of the calomel electrode
may get plugged with silver chloride.

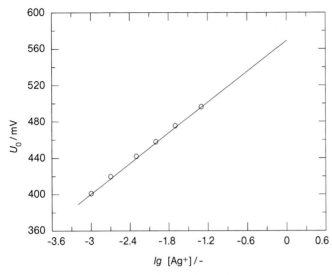

Fig. 2.4 Plot of the cell voltages obtained during the determination of the standard electrode potential of the silver electrode

- Why do you need cells without transference for exact determination of standard electrode potentials?
- Discuss the error caused by variations of hydrogen pressure (ambient atmospheric pressure).
- Which systematic error must be taken into account when using the method described here?
- Describe the design of the Weston cell. Why does it guarantee a constant cell voltage?

Experiment 2.3: pH-Measurements and Potentiometrically Indicated Titrations

Tasks
- Registration of a calibration curve for a glass electrode and an antimony electrode
- Calibration of a pH-meter with a glass electrode
- Potentiometrically indicated titration of formic acid, acetic acid, propionic acid, chloroacetic acid, and dichloroacetic acid with KOH; determination of the respective pK_a-values[7] from the titration curves.

Fundamentals
Of all pH-measurement methods the potentiometric method is the most important one. Starting with the definition of the pH-value

[7] The general concentration equilibrium constant K_c is modified and specified for acid dissociation K_a.

$$pH = -lg \, a_{H^+} \tag{2.32}$$

it is obvious that a pH-measurement is strictly speaking the determination of a single ion activity. Even with cells without diffusion potentials (see Exp. 2.2), only mean activity values can be obtained. They are equal only at very dilute concentrations (Debye-Hückel region). Thus, as a standard for pH-measurements, selected buffer solutions are used (IUPAC recommendation) with a hydrogen electrode as proton-selective electrode in a setup:

H$_2$-electrode/buffer solution/salt bridge 1 M KCl/reference electrode

The pH-value of the buffer solution is related to the measured cell voltage and the potential of the reference electrode E_{ref}:

$$pH = (U_0 - E_{ref})/(2.303 \cdot R \cdot T/F) \tag{2.33}$$

This way the conventional pH-scale is established, the basis of practical pH-measurements. With these standard buffers as reference, pH-values of solutions with unknown concentrations can be determined, and the dependency of the voltage of a cell with a pH-sensitive electrode on the pH-value can be obtained (so-called "electrode function").

Preferred pH-sensitive electrodes are the glass electrode, the quinhydrone electrode (EC:148), and some metal oxide electrodes (Sb- and Bi-electrode). The most popular electrode is the glass electrode. The most convenient constructions contain both electrodes (the pH-sensitive and the reference electrode) in a single cell body. These electrodes cannot be produced with exactly equal properties. In addition, sensitivity and zero-point (the voltage measured with solutions of equal pH inside and outside the device) are subject to ageing. Calibration is done with at least two different buffer solutions. A typical result is displayed below (Fig. 2.5).

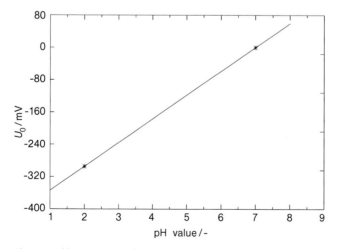

Fig. 2.5 Calibration curve of a glass electrode.

For exact measurements, the pH-values of the used buffers should be similar to the unknown ones. With the calibration, the slope in mV/pH-unit or (equivalent and more popular) the sensitivity in % (100% equals the theoretical value from the Nernst equation: 59 mV/pH-unit) and the zero-point (pH-value with voltage 0 expected with equal pH-values inside and outside) are adjusted. In addition the temperature must be taken into account.

Quantitative concentration measurements for analytical purposes (e.g., determination of the concentration of a solution of HCl) are not very precise because of the logarithmic relationship between activity and electrode potential. Potentiometry can nevertheless be used conveniently for indication of the equivalent point in titrations. The electrode used to detect changes in the titrated solution is called the indicator electrode; selection depends on the type of titration. In argentometry a silver electrode is used, in redox titrations a platinum or glassy carbon electrode may be appropriate; in acid-base titrations a glass electrode is suitable. The equivalence point is indicated by a large change in cell voltage (i.e. potential difference between indicator and reference electrode). A plot of cell voltage versus added volume of titration solution yields characteristic curves with a turning point (if necessary determined by drawing a tangent line) at the equivalence point. A typical case is shown below (Fig. 2.6).

During titrations of weak or medium-strong acids the equivalence point is not located at pH=0 because the salt formed as a product of neutralization is hydrolyzed. When 50% of the initially present acid is converted (i.e. neutralized), the titrated solution contains equimolar fractions of salt and acid, equivalent to a buffer solution with well-defined pH-value. From the pH-value corresponding to this point of the titration curve the dissociation constant K_a can be determined according to

$$HA \leftrightarrows H^+ + A^- \tag{2.34}$$

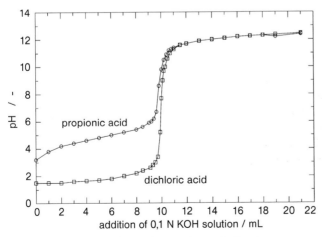

Fig. 2.6 Determination of equivalence points and pK_a-values.

$$K_a = (a_{H^+} a_{A^-})/a_{HA} \tag{2.35}$$

$$-\lg K_a = -\lg a_{H^+} - \lg a_{A^-}/a_{AH} \tag{2.36}$$

With and $a_{A^-} \cong c_{salt}$ and $a_{AH} \cong c_{HA} \cong c_{Acid}$ an approximation is possible [8]:

$$-\lg K_a = pH \tag{2.37}$$

Eq. (2.37) is valid only for weak acids because only in the case of very incomplete dissociation will the activity of the formed salt be approximately equivalent to the concentration $a_{A^-} \cong c_{salt}$ and the fraction of undissociated acid will be very close to the total acid concentration (and activity). The data plotted in Fig. 2.6 yield for dichloroacetic acid $K_a = 10^{-1.3}$; a literature value of $K_a = 10^{-1.29}$ has been reported (Handbook of Chemistry and Physics, 86th edition, 8–42). The value for propanoic acid is $K_a = 10^{-4.8}$, a literature value is $K_a = 1.4 \cdot 10^{-5}$ (P. W. Atkins, J. de Paula, Physical Chemistry, 8th ed., Oxford University Press, Oxford 2006, p. 1007).

Execution
Chemicals and Instruments
Standard buffer solutions pH = 9.18, 6.86, 4.01, and 1.68
Aqueous solutions (0.1 M) of acetic acid, formic acid, propionic acid, chloroacetic acid, and dichloroacetic acid
Aqueous titration solution of KOH (0.1 M)
pH-meter
Glass electrode
Antimony electrode
Saturated calomel electrode
Glass beakers 100 ml
Magnetic stirrer
Magnetic stirrer bar

Setup
During setup and execution, mechanical sensitivity of the thin membrane of the glass electrode must be kept in mind.

Recording calibration curves
- Wiring of the measurement circuit (connect glass electrode to pH-meter, install antimony electrode and reference electrode, and connect to pH-meter set to voltmeter mode).
- Put buffer solution into beaker, dip electrodes into solution, and record cell voltage when value is stable. Discard buffer solution or return it into storage vessel; do not return into original bottles.

Hint: In strongly acidic solution the antimony electrode is unstable.

---

8) The subscript HA indicates the undissociated fraction of acid;
acid refers to the total amount of acid added into the solution.

Calibration of pH-meter[9] with glass electrode
- Dip glass electrode into buffer solution pH=6.86.
- Set range to 0 ... 14.
- Adjust "sensitivity" to 100% (right limit).
- Adjust temperature setting to actual value in buffer solution.
- Adjust displayed pH value exactly to pH=6.86 with knob "sensitivity"[10].
- Rinse electrode carefully and exchange buffer solution:
- New solution pH=4.01 for measurements at pH<7.
- New solution pH=9.18 for measurements at pH>7.
- (Since following acids are investigated use pH=4.01).
- adjust to exactly pH=4.01 with knob "sensitivity".
- check with third buffer solution.

Potentiometric titration of weak acids
- The acids are supplied at concentrations $c=0.1$ **M**. Use the calibrated pH-meter with attached glass electrode.
- Put 10 ml of acid into beaker, add magnetic stirrer bar, place beaker on magnetic stirrer plate, adjust rate of rotation low enough to keep stirrer from hitting the glass membrane. Titrate by adding solution of KOH (0.1 **M**) in 1-ml steps.

Evaluation
- Plot the calibration curve and determine slope of curve in mV/pH-unit. Discuss electrode properties and estimate precision of measurement.
- Plot titration curve as shown in Fig 2.6. Determine equivalent point and pH-value at 50% conversion of acid.
- Calculate K_a-value and compare with literature data. Set up table of results for all acids and discuss relationship between strength of acid, type of intramolecular bondings, and deviation from literature data for chlorinated acids.

The following calibration curve (Fig. 2.7) was obtained with an antimony electrode as proton- (and pH)-sensitive electrode and a saturated calomel electrode as reference electrode in buffer solutions of various compositions.

The calculated sensitivity of 53 mV/pH-unit is somewhat below the theoretical value of 59 mV/pH-unit. The instability of the electrode at low pH-values is obvious.

---

9) The actual labels on the front plate of the pH-meter may vary; instead of asymmetry zero-point may be found.

10) If no proper adjustment is possible the glass electrode may be degenerated by, e.g., dryness, damage of membrane. If there is no visible damage or lack of solution in the glass electrode (some glass electrodes can be refilled) extended soaking in an aqueous solution of KCl (3 **M**) may regenerate the electrode. The instruction manual of the electrode may provide further help.

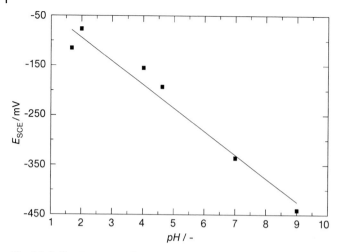

Fig. 2.7 Calibration curve of an antimony electrode as pH-sensitive electrode.

Questions
- Describe the establishment of the cell voltage of a glass electrode.
- What is a conventional pH-scale? Why was it established?
- Which electrode reactions are proceeding at a quinhydrone and at an antimony electrode?
- Describe the advantages of a potentiometric titration.
- What is a buffer solution? Describe representative acidic, neutral, and alkaline buffer systems.

Experiment 2.4: Redox Titrations (Cerimetry)

Task

The amount of Fe(II) ions in a sample solution shall be determined by redox titration with a solution of Ce(IV) ions and a platinum indicator electrode.

Fundamentals

Ce(IV) ions are strong oxidants ($E_{00,Ce^{3+}/Ce^{4+}} = +1.44 \text{ V}$[11]) and can be used as titrants ("cerimetry"). Because of their weak coloration, detection of the equivalence point based on color changes is not feasible. Potentiometric determination with an indicator electrode made of an inert electron conducting material like, e.g., platinum are possible. The observed titration curve shows an almost horizontal part at additions of titration solution before the equivalence point, a large

11) Values of E_{00} published in reference books vary considerably;
the value given here is confirmed by this experiment. The
value depends on the composition of the electrolyte solution
("real potential").

slope around the equivalent point, and again an almost horizontal line beyond it.

The potential of the indicator electrode is always determined by the concentrations (or more precisely the activities) of the participating redox ions at low degrees of titration (at the beginning) and at high degrees (considerable excess of titrant). In the first case the potential is solely controlled by the concentrations of the ions to be titrated because the titrant itself is present in one form only (the added form is consumed immediately and completely by the homogeneous redox reaction) and thus cannot determine a redox potential itself. At a degree of titration 0.5 (half-way to the equivalence point) the ratio of concentrations of the reduced and the oxidized form of the titrant is 1, and thus the concentration-dependent term in the Nernst equation vanishes. The electrode potential of the indicator electrode is equal to the standard potential. At a degree of titration 2 (much beyond the equivalent point), ions of the kind to be determined are present only in their converted form; they cannot establish a redox potential. The concentrations of the oxidized and the reduced form of the titrant are equal, and again the concentration-dependent term in the Nernst equation vanishes. Now the potential of the indicator electrode is equal to the standard potential of the titrant. Accordingly titration curves obtained during redox titrations can be used to determine further electrochemical data beyond the concentration.

Execution
Chemicals and instruments
Aqueous solution of Fe^{2+} (**0.01 M**)
Aqueous solution of Ce^{4+} (**0.01 M**)
Beaker
Burette
Saturated calomel reference electrode
Platinum electrode as indicator electrode
High input impedance voltmeter
Magnetic stirrer
Magnetic stirrer bar

Setup
The solution containing iron ions at unknown concentration is put in the beaker; pure water and the magnetic stirrer bar are added. The platinum and the calomel electrode are placed at a safe distance from the stirrer bar. The reference electrode is connected to the "low" input of the voltmeter.

Procedure
The cerium ion solution is added in small volumes (initially 0.5 ml; close to the equivalent point even smaller). Additions are stopped at a total volume equivalent to double the amount added at the equivalence point (degree of titration 2).

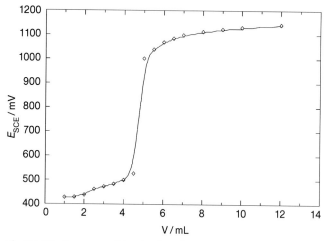

Fig. 2.8 Titration curve of the titration of Fe(II) ions (5 ml solution with $c=0.01$ **M**) with Ce(IV) ions ($c=0.01$ **M**).

Evaluation

A typical titration curve is shown in Fig. 2.8. At a degree of titration 0.5 the indicator electrode shows a potential $E_{SCE}=0.53$ V; the literature value for the Fe(II)/Fe(III) redox system is $E_{SCE}=0.53$ V. At a degree of titration 2 a potential of $E_{SCE}=1.12$ V is found; the literature value for the Ce(III)/Ce(IV) redox system is $E_{SCE}=1.2$ V (note the conversion from the value relative to the standard hydrogen electrode given above to the value on the saturated calomel scale). The observed deviations are mostly due to differences in composition of the electrolyte solution.

Experiment 2.5: Differential Potentiometric Titration

Task

Determination of the content of Fe(II) ions by redox titration with Ce(IV) ions by the differential potentiometric titration method and platinum indicator electrodes.

Fundamentals

Aqueous solutions of Ce(IV) ions are strong oxidants ($E_{00,Ce^{3+}/Ce^{4+}} = +1.44$ V), and because of their higher stability when compared with permanganate solutions they are preferred reagents in redox titrations ("cerimetry"). Their color is weak only; color changes are insufficient for detection of the equivalent point. A plot of the change of cell voltage (of the cell composed of the platinum indicator electrode and the reference electrode) observed after addition of a fixed volume of titration solution as a function of total volume yields a plot as shown below (Fig. 2.9):

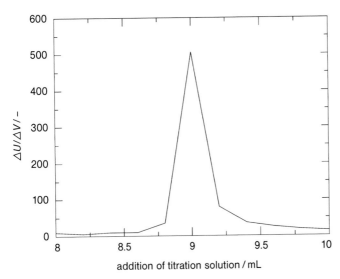

Fig. 2.9 $\Delta U/\Delta V = f(V)$ during a potentiometric titration.

The plot is equivalent to the first derivative of the corresponding titration curve showing the cell voltage as a function of the added volume of titrant (see Fig. 2.8). The maximum in Fig. 2.9 indicates the equivalence point. In a differential potentiometric titration this curve is obtained directly (i.e. without differentiation). Two indicator electrodes of the same material are dipped in the solution; one electrode is surrounded by a capillary. The capillary keeps changes in the titration solution caused by the addition of titrant away from this indicator electrode; this electrode maintains its potential. The cell voltage now corresponds to the concentration change effected by the addition; the cell is a "concentration cell" (EC: 86; 105). In this cell the cell voltage (electrode potential difference) is not generated by a chemical cell reaction, but by differences in concentration of the participating ions and the subsequent transfer of ions from places of higher activity to those of lower activity. With activities inside the capillary a' the cell voltage before reaching the equivalent point is given by:

$$U = E - E' = ((R \cdot T)/F) \cdot \ln((a_{Fe^{3+}} \cdot a'_{Fe^{2+}})/(a_{Fe^{2+}} \cdot a'_{Fe^{3+}})) \tag{2.38}$$

The cell voltage is solely caused by differences in activity; any terms related to the standard potentials of the electrode reactions vanish. If the solution inside the capillary is not exchanged during the whole titration a normal titration curve (see Fig. 2.8) is obtained. When the solution inside the capillary is mixed with the bulk solution after every addition of titrant the ratio $\Delta U/\Delta V$ is obtained directly. It shows a maximum at the equivalent point.

Execution

Chemicals and instruments

Aqueous solution of Fe^{2+}, 0.01 **M**

Aqueous solution of Ce^{4+}, 0.01 **M**

Cell for differential potentiometric titration with platinum electrodes

High input impedance voltmeter

Burette

Magnetic stirrer

Magnetic stirrer bar

Setup

The cell is shown schematically in Fig. 2.10. Exchange of solution in the capillary is effected by pressing the rubber bulb.

Procedure

- To 10 ml of a sample solution containing Fe(II) (c=0.01 **M**) in the beaker, water is added until capillary and indicator electrode are immersed. With the rubber bulb the capillary volume is purged and filled with sample solution.
- Connect electrodes to voltmeter, switch on voltmeter.
- Record data for the normal titration curve by adding titration solution without exchange of solution in the capillary volume. Start with additions of 1 ml, close to the equivalent point smaller volumes are added; later larger ones may be added.
- Record data for the differential potentiometric titration curve by exchanging the solution inside the capillary with the bulk in the beaker by pressing and releasing the rubber bulb carefully after every addition of titrant until the cell voltage reaches a minimum. Amounts off added volume are adjusted as before.
- Calculate the amount of Fe(II) in the initial sample.

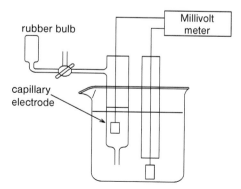

Fig. 2.10 Cell for differential potentiometric titration.

Evaluation

Plot both titration curves and determine equivalence points. Typical results are shown below (Fig. 2.11).

The advantage of this approach becomes immediately apparent on comparing this plot with the result obtained with the same setup without exchange of solution as displayed in Fig. 2.12.

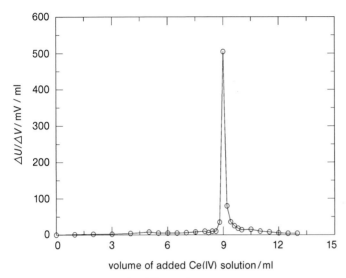

Fig. 2.11 Titration curve of a differential potentiometric titration with exchange of solution after every addition of titrant.

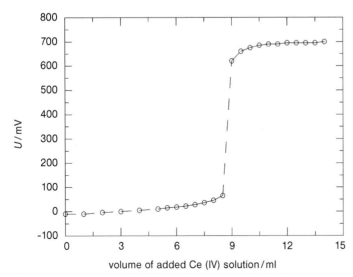

Fig. 2.12 Titration curve of a differential potentiometric titration without exchange of solution after every addition of titrant.

Questions

- Describe a concentration cell.
- Develop a setup for differential potentiometric titration with the argento-metric method.

Experiment 2.6: Potentiometric Measurement of the Kinetics of the Oxidation of Oxalic Acid

Task

The homogeneous oxidation reaction of oxalate with permanganate is monitored potentiometrically at different temperatures in order to obtain the reaction rate constant and energy of activation.

Fundamentals

The oxidation of oxalic acid proceeds according to

$$2\,MnO_4^- + 5\,C_2O_4^{2-} + 16\,H^+ \rightarrow 2\,Mn^{2+} + 8\,H_2O + 10\,CO_2 \tag{2.39}$$

This reaction was the first to be studied (by Harcourt in Oxford) from the point of view of the elucidation of the laws of chemical reaction kinetics. The choice was a rather unlucky one – the reaction is autocatalytic.

For a simple autocatalytic reaction

$$A \xrightarrow[B]{k} B \tag{2.40}$$

with initial concentration $c_{i,0}$ of substance I and – to simplify subsequent integration – the degree of conversion x (the fraction x of the total concentration of reactant converted at time t)

$$\frac{dx}{dt} = k \cdot (c_{A,0} - x) \cdot x \tag{2.41}$$

At $x = c_{A,0}/2$ the rate is at a maximum. At $x=0$ the rate is zero. Assuming a finite – albeit small – initial concentration of B ($c_{B,0}$) eq. (2.41) is modified into

$$\frac{dx}{dt} = k_1 \cdot (c_{A,0} - x) \cdot (c_{B,0} + x) \tag{2.42}$$

Double integration from 0 to t and 0 to x yields

$$\frac{1}{c_{A,0} + c_{B,0}} \ln\left(\frac{c_{A,0} \cdot (c_{B,0} - x)}{c_{B,0} \cdot (c_{A,0} - x)}\right) = k \cdot t \tag{2.43}$$

With $c_0 - c_t = x$ simplification is possible

$$k = \frac{1}{t \cdot (c_{A,0} + c_{B,0})} \ln \frac{c_{A,0} \cdot c_{B,t}}{c_{B,0} \cdot c_{A,t}} \tag{2.44}$$

Observation of color changes enables simple examination of the claimed autocatalysis. During the application of this reaction in quantitative analysis in measurements of oxalic acid, after the initial addition of permanganate solution no decoloration is observed; actually one might be afraid of having added too much already. After several seconds discoloration proceeds nevertheless, and subsequently discoloration occurs rapidly after every addition of permanaganate solution until the equivalence point is reached. The Mn^{2+} ions formed in the titration reaction are acting as catalysts (autocatalysis).

The progress of the reaction can be monitored visually quite easily – disappearance of the pink color very strong; even at low permanganate concentrations conversion with respect to the added permanganate can be assumed to be completed. More precisely it can be stated only that conversion has proceeded to a state where the (low) concentration of permanganate cannot be observed visually any more. Instead, the reaction can be monitored potentiometrically by measuring the redox potential of the electrode MnO_4^-/Mn^{2+}. The redox electrode potential is given by

$$E_0 = E_{00} + \frac{R \cdot T}{z \cdot F} \ln \frac{a_{ox} a_{H^+}^8}{a_{red}} \tag{2.45}$$

In the evaluation of the electrode potential vs. time curves we study the turning point (at which visually observed permanganate is apparently completely converted). We identify the educt A in Eq. 2.40 with the permanganate; the manganese ion refers to B. A change of the electrode potential from the initial value E_i to a value E_t at the turning point can be correlated with the concentration of permanganate and manganese ions and the respective concentration ratios:

$$\Delta E = E_i - E_t = E_{00} - \frac{R \cdot T}{z \cdot F} \ln \frac{a_{ox,i} a_{H^+,t}^8}{a_{red,i}} - E_{00} - \frac{R \cdot T}{z \cdot F} \ln \frac{a_{ox,t} a_{H^+,i}^8}{a_{red,t}} \tag{2.46}$$

Assuming a constant pH-value fixed by the added sulfuric acid and activities being approximately equal to concentrations at the small values employed here, simplification yields

$$\Delta E = \frac{R \cdot T}{z \cdot F} (\ln(c_{ox,i}/c_{red,i}) - \ln(c_{ox,t}/c_{red,t})) \tag{2.47}$$

or

$$\Delta E = \frac{R \cdot T}{z \cdot F} \ln \frac{c_{ox,i} \cdot c_{red,t}}{c_{ox,t} \cdot c_{red,i}} \tag{2.48}$$

Rearrangement results in

$$\ln\frac{c_{ox,i} \cdot c_{red,t}}{c_{ox,t} \cdot c_{red,i}} = \frac{\Delta E \cdot z \cdot F}{R \cdot T} \tag{2.49}$$

This equation can be inserted in Eq. (2.44); k is obtained as

$$k = \frac{\Delta E \cdot z \cdot F}{R \cdot T \cdot t_{tp}(c_{ox,i} + c_{red,i})} \tag{2.50}$$

with time t_{tp} at the turning point of the potential versus time plot. The term $(c_{ox,i} + c_{red,i})$ is equal to the initially added concentration of permanganate.

Execution
Chemicals and instruments
Aqueous solution of sulfuric acid 0.01 **M**
Aqueous solution of $KMnO_4$ 0.01 **M**
Aqueous solution of oxalic acid 0.05 **M**
Platinum electrode
Saturated calomel electrode
High input impedance voltmeter
Thermostat
Double-walled measurement cell (thermostat jacket)
Stop watch
Pipette 5 ml
Graded cylinder 100 ml

Setup
Electrodes are connected to the voltmeter (calomel electrode to "low", platinum electrode to "high" input); the thermostat jacket is connected to the thermostat.

Procedure
Into the cell thermostatted at 35 °C 100 ml water, 5 ml sulfuric acid and 5 ml permanganate solution are added. If everything works well a cell voltage of about 0.9 V will be observed. The oxalic acid is added with the pipette; when half the complete volume has been added the stopwatch is started. The initially small potential changes are recorded at longer time intervals; around the turning points the time intervals should be smaller, and after the potential drop only a few more values should be recorded. The experiment is repeated at higher temperatures when values should be recorded at very short time intervals.

Evaluation
Fig. 2.13 shows typical potential versus time plots at various reaction temperatures. From the time t_{tp} up to the turning point the value of k can be calculated according to

$$(\Delta E \cdot z \cdot F)/(R \cdot T) = \ln(c_{ox,i}/c_{ox,t}) = k \cdot t \tag{2.51}$$

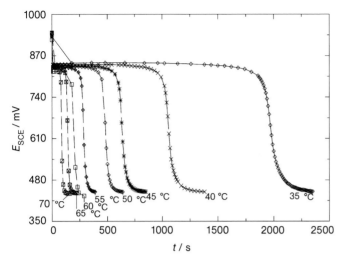

Fig. 2.13 Potential versus time curves recorded during oxidation of oxalic acid with KMnO$_4$ in aqueous acid solution.

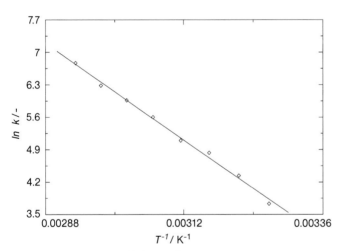

Fig. 2.14 Arrhenius plot of ln k vs. T^{-1}.

From the plotted data the following rate constants were obtained:

$T/°C$	$k/l·mol^{-1}·s^{-1}$	$T/°C$	$k/l·mol^{-1}·s^{-1}$
35	41.7	55	270.7
40	76.6	60	389.6
45	125.3	65	534.7
50	163.2	70	867.8

From the Arrhenius plot (Fig. 2.14) the activation energy of $E_a = 72$ kJ·mol^{-1} is calculated.

Literature
S. R. Logan: Fundamentals of chemical kinetics, Longman, Essex 1996.

Experiment 2.7: Polarization and Decomposition Voltage [12]

Task
The temperature dependence of the decomposition voltage of an aqueous solution of HCl (1.2 **M**) and the concentration dependence of the decomposition voltages of HBr and HI are to be determined.

Fundamentals
The equilibrium voltage U_0 of a galvanic or an electrolysis cell calculated from thermodynamic data can be examined experimentally in a static method (see Exp. 2.2) or in a dynamic one. In the second approach the voltage applied to an electrolysis cell is raised slowly; a plot of the measured current versus the applied voltage yields from extrapolation to $I \rightarrow 0$ the decomposition voltage U_d equivalent to U_0. Differences between experiment and calculation are caused by slow electrode reaction (polarization or overvoltage [13]); in the halogen evolution reaction studied here these deviations are not significant.

The relation between the Gibbs energy of a spontaneously proceeding reaction and the cell voltage is given by

$$\Delta G = -z \cdot F \cdot U_0 \tag{2.52}$$

Because electrolysis proceeds only when an external voltage is applied, i.e. at $\Delta G > 0$, the relationship between ΔG and U_0 is

$$\Delta G = z \cdot F \cdot U_0 \tag{2.53}$$

With the Gibbs equation the Gibbs energy of reaction ΔG and after determination of the temperature dependence of U_0 the entropy of reaction can be obtained.

---

12) Because a current – although a small one only – is flowing in this experiment, strictly speaking it belongs in the following chapter. Because the aim is the determination of thermodynamic data at equilibrium by extrapolation to current zero the experiment seems to be better placed here. The term decomposition voltage is a linguistic compromise. It refers to the minimum cell voltage needed for decomposition of the electrolyte, i.e. deposition of electrolysis products. Unfortunately the term deposition voltage does not seem to be any better.

13) Overvoltage ΔU is the difference between the cell voltage at current zero and at a finite current $\Delta U = U - U_0$. With respect to a single electrode the term overpotential with its own symbol η is defined in the same manner: $\Delta E = E - E_0$.

Execution
Chemicals and instruments
Aqueous solution of HCl 1.2 **M**
Potassium bromide
Potassium iodide
Hydrogen
Platinum tip electrode
Platinized platinum wire net electrode (used as hydrogen electrode)
Adjustable voltage source
Microammeter
Thermostat
Double-walled measurement cell (thermostat jacket)

Setup and Procedure
- In the electrochemical cell connected to the thermostat and equipped with electrodes as depicted schematically below (Fig. 2.15), U_d for HCl is measured at standard conditions ($a_{Cl^-} = 1$ **M**, $p = 1$ atm ($= 101\,325$ Pa, $a_{H^+} = 1$).
- 75 ml of hydrochloric acid solution (1.2 **M**) are placed in the cell, the hydrogen stream is adjusted to a constant slow flow of bubbles around the platinum net electrode, the cell temperature is adjusted to 15 °C.
- The experiment is started at $U = 0$ V. The voltage is slowly increased (Please explain the current oscillations observed on the microammeter). When a voltage of $U = 1.2$ V is reached the increase is stopped. After the current has dropped to zero the actual measurement starts. The voltage is raised in steps of 0.02 V, current is recorded after one minute. When the current approaches 100 µA the experiment is finished.
- The measurement is repeated at $T = 25$ °C, 35 °C and 45 °C.
- U_d for 0.1 **M** and 1 **M** solutions of HBr and HI are determined. To 75 ml of HCl solution 0.9 g KBr are added (this results in a solution 0.1 **M** in HBr). After completing the measurement further 8.1 g KBr are added (final concentration 1 **M** HBr). Again the current-potential relationship is measured. To a

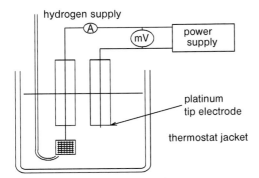

hydrogen supply

platinum
tip electrode

thermostat jacket

Fig. 2.15 Experimental setup for decomposition voltage measurement.

fresh solution of HCl (1.2 **M**) first 1.26 KI (resulting in 0.1 **M** HI) and after
the measurement further 11.34 g KI (final concentration 1 M HI) are added.
As before the actual measurement starts at 0.4 V with HBr and at 0.8 V with
KI. If currents at these voltages are too large, recordings should be started at
lower initial values. Above these voltages again currents are recorded after
voltage increases in small steps (0.02 V).

Evaluation

The obtained current-potential relationships are plotted as shown in a typical example below (Fig. 2.16).

From U_d obtained by extrapolation to zero current the Gibbs energy of HCl
electrolysis is calculated assuming $U_d = U_0$ at zero current and compared with
literature data according to

$$\Delta G = z \cdot F \cdot U_0 \tag{2.54}$$

The Gibbs energy at $T = 288$ K is $\Delta G = 133.92$ kJ·mol^{-1}. The respective value at
$T = 298$ K is $\Delta G = 132.1$ kJ·mol^{-1}. The corresponding literature value is $\Delta G = 131.23$ kJ·mol^{-1} (P.W. Atkins, J. de Paula, Physical Chemistry, 8th ed., Oxford
University Press, Oxford 2006, p. 996).

From the values of U_d obtained at elevated temperatures the reaction entropy
is calculated from $\partial U_d / \partial T$. For comparison reaction entropies should be calculated from literature data and be compared with the experimentally obtained
ones. The reaction enthalpy ΔH is assumed to be constant in the studied range
of temperatures. Figure 2.17 shows typical data for electrolysis of HCl.

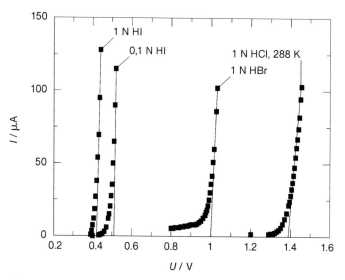

Fig. 2.16 Current-potential curves with various halide solutions at $T = 288$ K.

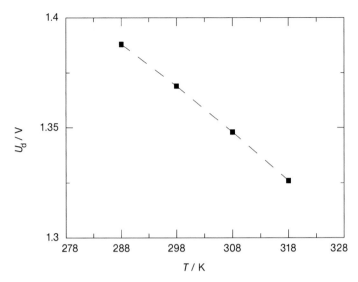

Fig. 2.17 Plot of U_d for an aqueous solution of HCl 1.2 **M**.

From the temperature coefficient $\partial U_d/\partial T = -2 \text{ mV} \cdot \text{K}^{-1}$ of U_d (which is assumed to be equal to U_0) the entropy of the electrolysis reaction is calculated according to

$$\Delta S = -\partial U_d/\partial T \cdot z \cdot F = 0.002 \cdot z \cdot F = 192.97 \text{ J} \cdot \text{K}^{-1} \cdot \text{mol}^{-1} \tag{2.54}$$

This result differs from the value $\Delta S = 120.4 \text{ J} \cdot \text{K}^{-1} \cdot \text{mol}^{-1}$ calculated from literature data (EC:89, P.W. Atkins, J. de Paula, Physical Chemistry, 8th ed., Oxford University Press, Oxford 2006, p. 996).

Appendix
In many cases the experimentally determined values of U_d differ substantially from those calculated from thermodynamic data. As an example the calculated value for water electrolysis is $U_d = 1.229$ V. In a real cell even with platinum electrodes and water acidified with sulphuric acid voltages of 1.6...1.8 V are needed. The electrode reactions are

$$2 \text{ H}^+ + 2 \text{ e}^- \rightarrow \text{H}_2 \tag{2.55}$$

$$\text{H}_2\text{O} \rightarrow {}^{1}\!/_{2}\text{O}_2 + 2 \text{ e}^- \tag{2.56}$$

The observed differences are associated with (as well as other, minor causes) the electron transfer reactions at the phase boundary metal/solution. In the case of unimpeded charge transfer, i.e. of a very fast electron transfer reaction), a current-potential relationship as shown in Fig. 2.16 with a measurable flow of current already at the decomposition voltage $U_d = 1.229$ V calculated from thermodynamic data would have been expected. If there is any hindrance of charge transfer at one or both electrodes, a more or less considerable additional voltage (overvoltage) $\eta = U - U_d$ will be needed. In this case evaluation of the current-potential relationship will not result in useful thermodynamic data.

Investigations of the behavior of electrodes with flowing current are the subject of electrode kinetic studies. It has been observed that the chlorine evolution reaction

$$2 \ Cl^- \rightarrow Cl_2 + 2 \ e^- \tag{2.57}$$

proceeds almost unimpeded. The reaction

$$H_2O \rightarrow 2 \ H^+ + \frac{1}{2} \ O_2 + 2 \ e^- \tag{2.58}$$

is strongly hindered whereas the hydrogen evolution reaction HER proceeds almost unimpeded:

$$2 \ H^+ + 2 \ e^- \rightarrow H_2 \tag{2.55}$$

The experiment described above is possible only because of the slow oxygen evolution. Otherwise, and from a thermodynamic point of view, oxygen evolution will start before chlorine evolution.

The rate of an electrode reaction (whether it is impeded or accelerated) depends on the electrode material. Electrolysis of hydrochloric acid at a hydrogen-generating mercury electrode and a platinum sheet for chlorine evolution results in an experimental value of $U_d = 2$ V. Hydrogen evolution is strongly impeded at the mercury electrode.

In order to avoid unwanted contributions to the observed cell voltage from the already small hindrance of hydrogen evolution kinetics, the experiment is not done with smooth platinum sheets of approximately equal size with one electrode exposed to a stream of hydrogen bubbles. Instead, a platinized platinum wire net electrode with relatively large effective surface area is used as the hydrogen electrode whereas a smooth platinum tip with only a few mm^2 surface area is used as the chlorine electrode. Because of this small surface area only very small currents will flow (up to approx. 50 A). These currents do not affect the electrode potential of the large-area hydrogen electrode. More important than current is the current density j ($j = I/A$). Overpotentials increase with growing j. In this experiment the obtained current-potential curve is predominantly determined by the reaction at the chlorine electrode because of its relatively small surface area and thus larger j.

A chlorine electrode

$$2 \ Cl^- \leftrightarrows Cl_2 + 2 \ e^- \tag{2.57}$$

is composed of a platinum electrode in contact with an aqueous solution of HCl exposed to a stream of chlorine gas. At standard conditions ($a = 1$ M (approx. $c_{HCl} = 1$ M), $p = p_0 = 1$ atm (= 101 325 Pa)) the potential difference with respect to a standard hydrogen electrode according to

$$2 \ H^+ + 2 \ e^- \leftrightarrows H_2 \tag{2.55}$$

can be calculated from the thermodynamic data of the reaction

$$2 \ Cl^- + 2 \ H^+ \leftrightarrows Cl_2 + H_2 \tag{2.57}$$

as $U_d = 1.37$ V.

Literature

A. J. Bard and L. R. Faulkner: Electrochemical Methods, Wiley, New York 2001.

Questions

- Define the terms "overpotential" and "overvoltage".
- Describe the process from cell voltage measurements to Gibbs energies and reaction entropies.
- Name the differences between static and dynamic measurements of equilibrium cell voltages.
- Explain the current potential curve for water and hydrochloric acid electrolysis.

3
Electrochemistry with Flowing Current

Whereas in the previous chapter any flow of current was carefully avoided and precautions were observed to keep the system under investigation at thermodynamic-chemical equilibrium, in this chapter processes and phenomena are studied where a current flows resulting in observable, significant changes. First, the transport of charge and matter in electrolyte solutions is investigated. Possibilities of employing these observations in other areas of science (e.g., in analytical chemistry or process control) will be discussed in the next chapter. Experiments with molten or solid electrolytes are not included (although there are certainly many very appropriate ones around) because the experimental setups tend to be complicated and expensive and/or components are hard to get.

Following these experiments dealing with processes in the bulk of the electrolyte solution, measurements dealing with processes at the phase boundary between electronically conducting matter (e.g. metal) and ionically conducting matter (electrolyte solution) are described. Because some of the observed processes and phenomena will also show up in subsequent chapters, assignment of experiments to the present chapter turned out to be difficult. Finally an arbitrary assignment was attempted: Experiments dealing with fundamental aspects are collected in this chapter, and those dealing with applied aspects are assigned to the respective following chapters. Organizing experiments based on polarography was particularly difficult. Now all these experiments can be found in Chapter 4.

Electrochemical processes can be grouped according to the variable controlled by the operator into potentiostatic ones (where the electrode potential is set), galvanostatic (controlled current), and coulostatic (controlled charge) ones. As an alternative, organization according to the measured variable is possible: Potentiometric procedures (where the electrode potential is measured), voltammetric (potential is measured as a function of current, but also current as a function of electrode potential), amperometric (measurement of current), conductometric (measurement of electrolytic conductance), and coulometric (measurement of the consumed charge) ones. Neither scheme is perfect, and the following procedures and experiments are therefore initially grouped into those where the studied phenomenon is present in the bulk of the electrolyte solution volume and those where the process at the electrode/solution interface dominates. In this second part, classification according to the measured variable is

Experimental Electrochemistry. A Laboratory Textbook. Rudolf Holze
Copyright © 2009 WILEY-VCH Verlag GmbH & Co. KGaA, Weinheim
ISBN: 978-3-527-31098-2

observed. Some processes are preferably applied in analytical chemistry, and these are treated in the next chapter.

Experiment 3.1: Ion Movement in an Electric Field

Task
The associated movements of ions in a gel-like electrolyte are studied.

Fundamentals
In an electric field, charged particles are accelerated and put into motion towards one of the two electrodes establishing the field, with the direction depending on the sign of the charge on the particle. This accelerating force is counteracted by a braking force caused by the viscosity of the medium in which the particles are traveling. The resulting velocity v established when the accelerating and braking effects are at equilibrium is called the drift velocity. For an ion with radius r_i it is

$$v = (z \cdot e_0 \cdot E)/(6 \cdot \pi \cdot \eta \cdot r_i) \tag{3.1}$$

This movement can be observed easily with colored ions or with ions whose presence can be detected with colored indicators. The relationship between the velocity v of an ion i with number of charges z_i and the effective electric field is called the ionic mobility u_i

$$v = u_i \cdot E \tag{3.2}$$

or rearranged and more explicitly

$$u_i = v \cdot z_i/E = (z_i \cdot e_0)/(6 \cdot \pi \cdot \eta \cdot r_i) \tag{3.3}$$

Because the mobility of an ion depends on several parameters, which are in part substance-specific, this movement can be employed analytically for the separation of ions. In a procedure called electrophoresis the ion-containing substance to be analyzed is deposited on a carrier which is the viscous transport medium itself or contains this medium. In the present experiment filter paper is the carrier, and water soaked into the paper is the transport medium. In gel electrophoresis a cross-linked polyacrylamide gel is both carrier and transport medium. The medium is spread into a rectangular shape, and on two adjacent edges electrodes made from an inert material are attached. A drop of the analyte solution is deposited at a marked starting point. After applying the electric voltage (this may range from a few volts when very mobile ions and highly conducting media are employed to up to several kilovolts with poorly conducting media and ions of low mobility) the ions start to travel. Colored ions can be observed easily with the naked eye; the location of other ions during and after separation can be made visible with suitable indicators.

Execution

Chemicals and instruments
Agar
Potassium chloride
Diluted aqueous solution of KOH
Diluted aqueous solution of HCl
Diluted aqueous solution of $CuCl_2$
Alcoholic solution of phenolphthalein
U-shaped glass tube with perforated rubber stopper
2 carbon rod electrodes
Power supply

Setup
Agar (0.5 wt%) is mixed with warm water and some added KCl. The solution is filled into the U-shaped glass tube up to 2/3 of the tube height. Shortly before setting, some dilute KOH-solution with added alcoholic solution of phenolphthalein is stirred into the agar solution in the compartment later used for the anode. After stirring, the whole tube should show the typical red color of the indicator. Into the other tube dilute hydrochloric acid with some added indicator is stirred. After setting some dilute hydrochloric acid and a dilute solution of $CuCl_2$ are added into the former compartment, while into the latter one an equivalent amount of diluted KOH solution is added. Into both tubes the rubber stoppers with the carbon rod electrodes are fitted; because of the expected gas evolution the stoppers should not be fitted too tighthly. A DC voltage from the power supply is connected. The setup is illustrated in Fig. 3.1.

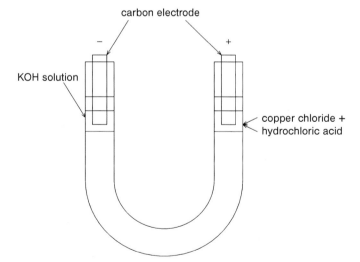

Fig. 3.1 Setup for the demonstration of the movement of ions and the difference in the mobility of ions.

Procedure

A few minutes after applying the DC voltage (the actual value depends on the conductance of the agar gel) gas evolution starts at both electrodes. In addition, changes of color can be observed. Copper ions move away from the anode into the gel. The initial red color of the gel close to the anode slowly disappears, and this extends further into the gel. In the other tube the color turns red, typical of the employed indicator in an alkaline environment.

Evaluation

Away from the anode, protons (of the hydrochloric acid) and copper ions move into the gel The progress of the former is indicated by the disappearance of the red color of the pH-indicator, and the presence of the copper ions is revealed by their blue color. The protons move faster because the boundary of the red color moves faster than that of the blue color. Hydroxyl ions move away from the cathode, as again indicated by the appearance of the red color of the pH-indicator. A comparison of the movement of the various color boundaries after some time $(1 \ldots 2 \text{ h})$ correlates well with the different ion mobilities (H^+: $36.23 \cdot 10^{-4}$ $cm^2 \cdot s^{-1} \cdot V^{-1}$, OH^-: $20.64 \cdot 10^{-4}$ $cm^2 \cdot s^{-1} \cdot V^{-1}$, Cu^{2+}: $5 \cdot 10^{-4}$ $cm^2 \cdot s^{-1} \cdot V^{-1}$).

Experiment 3.2: Paper Electrophoresis

Task

The movement of permanganate ions in a simplified paper electrophoresis arrangement is observed.

Fundamentals

In an electric field, charged particles move depending on their charge with respect to one of the electrodes. This process (for further details see Expt. 3.1) can be employed analytically. In an arrangement basically similar to those employed in paper and thin layer chromatography in a stationary phase (filer paper, gel), an ionically conducting phase (electrolyte solution) is fixed. At the short edges of the band-shaped strip, electrodes made from an inert material are attached. An electrical voltage is applied to these electrodes (the actual value depends on the conductance of the stationary medium and the fixed solution and ranges from a few volts to several kilovolts) and start ion movement. The distance covered by the ions varies depending on their mobility. Their progress is visible with colored ions or can be demonstrated with suitable indicators.

Execution

Chemicals and instruments
Small crystals of potassium permanganate
Aqueous solution of potassium chloride (0.01 **M**)
Microscopy glass slide (27·57 mm)
Filter paper
Aluminum foil
Power supply

Setup
On the short edges of the glass slide, aluminum foil is attached with adhesive tape leaving contact surfaces for a strip of filter paper placed on the glass slide and the contact strips. The paper is soaked with the potassium chloride solution, and now sticks to glass and foil. A DC voltage of approx. 7 V is applied. In the middle, between the electrodes, a crystal of $KMnO_4$ is placed on the wet paper.

Procedure
Initially traces of $KMnO_4$ will dissolve and color the immediate surroundings of the deep violet crystal[1]. Very soon the preferential movement of permanganate ions towards the anode becomes visible. The brown coloration of the paper is caused by oxidation of paper components resulting in formation of MnO_2.

Experiment 3.3: Charge Transport in Electrolyte Solution

Tasks
- Determination of the cell constant C of a conductance measurement cell.
- Determination of the specific conductance κ as a function of the concentration and of the equivalent conductance at infinite dilution Λ_0 with solutions of HCl, NaCl, CH_3COOH and CH_3COONa.
- Calculation of the degree of dissociation a of acetic acid as a function of concentration, calculation of the dissociation constant K_c.
- Determination of the saturation concentration of aqueous solution of $CaSO_4$ from conductance measurements.

Fundamentals
Salts, acids, and caustics (electrolytes) dissociate in sufficiently polar solvents into solvated ions, which are mobile carriers of electric charge and cause ionic

1) In a variation of this experiment a film of water is placed between the electrodes by carefully dripping the KCl-solution onto the glass slide. The crystal is placed in the middle of this film. Rapid diffusion of ions proceeds in all directions as well as the ionic migration towards the electrodes; in addition, the arrangement is labile and sensitive to mechanical shocks.

conductance L of the solution. The specific conductance κ (or the reciprocal specific resistance ρ) measured between two electrodes of area exactly $A=1$ cm^2 placed in a distance apart of $d=1$ cm in a solution depends on several experimental conditions and material properties. In order to compare different electrolytes, instead of the specific conductance

$$\kappa = L \cdot d/A \tag{3.4}$$

molar conductivities $\Lambda_{mol} = \kappa/c$ are used. Division by the ionic charge number z according to $\Lambda_{eq} = \kappa/(c \cdot z)$ results in the equivalent conductance.

Because a conductance measurement cell rarely has the exact dimensions as stated above, actual cells are calibrated by measuring the conductance of a KCl-solution of known conductance κ_{lit}. The cell constant C is calculated according to $C = \kappa_{lit}/L_{meas}$ (this is equivalent to the ratio d/A of the actual cell). In all following measurements, obtained conductivities L are corrected by multiplying by C.

At small concentrations, calculation of specific conductivities using the cell constant C must take into account the conductance of water κ_{water}. The amended equation for the specific conductance is

$$\kappa = (L_{sol} - L_{wat}) \cdot C \tag{3.5}$$

At higher concentrations κ_{water} can be neglected. The equivalent conductivities Λ_{eq} are plotted vs. $c^{1/2}$. From the plot the equivalent conductance at infinite dilution Λ_0 (extrapolation towards the y-axis) and the Kohlrausch constant k can be determined. The units of k can be derived from Kohlrausch's square root law

$$\Lambda_{eq} = \Lambda_0 - k \cdot c^{1/2} \tag{3.6}$$

or

$$k = \frac{d\Lambda_{eq}}{d\sqrt{c}} \tag{3.7}$$

This law does not apply to weak electrolytes (incompletely dissociated), in their case the conductance depends on the concentration-dependent degree of dissociation a associated with Λ_{eq} and Λ_0 according to

$$a = \frac{\Lambda_{eq}}{\Lambda_0} \tag{3.8}$$

Taking acetic acid as an example the concentrations of the various participating species are related according to $c_{H^+} = c_{Ac^-} = a \cdot c_0$, $c_{HAc} = (1-a)c_0 = c_{HAc,undiss}$ with the total concentration of acetic acid c_0. The dissociation constant K_c of Ostwald's dilution law can thus be calculated

$$K_c = \frac{a^2 \cdot c_0^2}{(1-a) \cdot c_0} = \frac{a^2 \cdot c_0}{1-a} \tag{3.9}$$

During studies of weak electrolytes it can be observed that plots of Λ_{eq} vs. $c^{1/2}$ do not yield straight lines. This is caused by the concentration-dependent degree of dissociation a. Extrapolation to obtain Λ_0 is uncertain, and practically almost impossible. Because Λ_0 is composed additively of cation and anion conductivities, the value of Λ_{0,CH_3COOH} can be calculated according to

$$\Lambda_{0,HCl} = \lambda_{0,H^+} + \lambda_{0,Cl^-} \tag{3.10}$$

$$\Lambda_{0,NaCl} = \lambda_{0,Na^+} + \lambda_{0,Cl^-} \tag{3.11}$$

$$\Lambda_{0,CH_3COONa} = \lambda_{0,Na^+} + \lambda_{0,CH_3COO^-} \tag{3.12}$$

Eq. (3.10) − Eq. (3.11) + Eq. (3.12)

$$\Lambda_{0,HCl} - \Lambda_{0,NaCl} + \Lambda_{0,CH_3COONa} = \lambda_{0,H^+} + \lambda_{0,CH_3COO^-} = \Lambda_{0,CH_3COOH} \tag{3.13}$$

Execution

Chemicals and instruments
Aqueous solutions of 0.1 **M** HCl, NaCl, KCl, CH_3COOH, and CH_3COONa
Saturated aqueous solution of $CaSO_4$
Conductance measurement cell
RCL-measurement bridge or conductance meter

Setup
For measurements of the cell resistance an RCL-bridge can be used; instead a conductance meter can also be used. In the latter case conversion from resistance into conductance values is not necessary.

With an RCL bridge all measurements are done at 1000 Hz. The temperature of the sample solution must be measured because the electrolytic conductance depends strongly on the temperature. During the measurement equilibration of the bridge (minimum reading at the indicator instrument) is first tried at low sensitivity; this procedure is repeated at higher sensitivity in order to obtain the cell resistance as precisely as possible.

Procedure
1. Determination of the cell constant C and the conductance of pure water

The measurement cell is rinsed with water until a stable resistance[2] value is obtained. The resistance of a solution of 0.01 **M** KCl is determined. Conductivities L are calculated as $L = 1/R$. From literature values of κ_{KCl} of the solution of KCl (if no literature value for the recorded temperature in the experiment is available interpolation may be necessary) the cell constant is determined according to

$$C = \kappa_{KCl}/L_{meas} \tag{3.14}$$

---

2) In the following description use of an RCL-bridge is assumed.

2. Measurement of concentration-dependent conductivities

The specific conductivities of solutions of the following electrolytes are determined:

2.1 HCl, NaCl, NaAc

$$c = 10^{-2}, 5 \cdot 10^{-3}, 10^{-3}, 5 \cdot 10^{-4}, 10^{-4} \text{ M}$$

2.2 HAc

$$c = 10^{-1}, 5 \cdot 10^{-2}, 10^{-2}, 10^{-3}, 10^{-4} \text{ M}$$

Solutions are prepared by dilution of the stock solution. Measurements are started with the most diluted solution.

2.3 Saturated aqueous solution of CaSO$_4$, diluted tenfold

The protocol contains measured conductivities, concentrations, and temperatures.

Evaluation

The protocol contains the calculated cell constant C and a plot of the determined equivalent conductivities Λ_{eq} vs. $c^{1/2}$ (Kohlrausch square root law, Fig. 3.2). By extrapolation Λ_0 is determined. With acetic acid this attempt fails, and an alternative route is described above.

From the conductance of the solution of CaSO$_4$ the solubility is calculated taking into account the tenfold dilution. In a typical experiment a value of $\kappa = 2.99 \cdot 10^{-4}$ S·cm^{-1} was found. With the conductance of pure water and the value of a concentration of CaSO$_4$ $c = 5.37 \cdot 10^{-4}$ mol·l^{-1} of the diluted sample is calculated; the saturation concentration of CaSO$_4$ is accordingly $c = 5.37 \cdot 10^{-3}$ mol·l^{-1}.

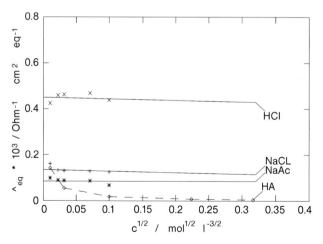

Fig. 3.2 Typical values of equivalent conductivities plotted according to Kohlrausch.

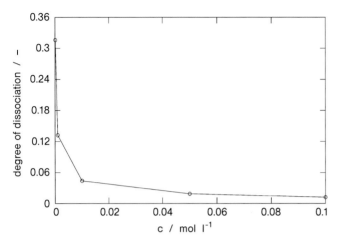

Fig. 3.3 Degree of dissociation α as a function of concentration.

From the solubility product $L = 2.4 \cdot 10^{-5}$ (Handbook of Chemistry and Physics, 86th edition, 8–118: $L = 4.93 \cdot 10^{-5}$) a concentration $c = 4.9 \cdot 10^{-3}$ mol·l^{-1} is calculated.

Questions
- Why is AC current used in conductance measurements?
- Why does a plot of specific conductance as a function of concentration pass through a maximum for most electrolytes?
- How large is the conductance of demineralized water and of ultrapure water? How can it be calculated from the ionic product of water?
- Must this conductance be taken into account in the described experiments? If yes, how?
- How can you explain the temperature dependence of electrolytic conductance?
- Can you separate the observed values Λ_0 into contributions of cations and anions? If yes, how? If no, what additional information do you need?
- Is the measured conductance influenced by stirring?
- Under what conditions does an electrolyte solution (i.e. the conductance measurement cell) behave like an Ohmic resistor?

Literature
M. R. Wright: An Introduction to Aqueous Electrolyte Solutions, John Wiley & Sons, Chichester 2007.

Experiment 3.4: Conductance Titration

Task
With conductometrically indicated titration the composition of various solutions of strong and weak electrolytes is determined.

Fundamentals
The conductance of an electrolyte solution changes as a function of the concentration of mobile charge-carrying ions. When this concentration changes during a titration because of, e.g., the chemical reaction between one solution constituent and the titrant, the conductance will change accordingly. This change can be employed to detect the equivalence point in a titration (EC:73). Because numerous titrants are electrolyte solutions themselves with respective contributions to electrolytic conductance, this approach does not look very attractive at first glance. This suddenly changes when the ionic equivalent conductivities $\lambda_0^\pm$ of the participating ions differ substantially. In case of an acid-base titration the highly conductive protons are consumed by neutralization and are replaced by cations of considerably smaller conductance. Beyond the equivalence point an excess of highly conducting hydroxyl ions is added, and conductance of solution increases rapidly again. In the V-shaped plot of conductance vs. volume of added titrant solution (alkaline solution in this example) the equivalence point can be identified easily at the minimum of the curve.

Execution
Chemicals and instruments
Aqueous solutions of 1 **M** HCl, KOH, CH_3COOH and CH_3COONa
Automatic burettes
Beaker 250 ml
Pipette 25 ml
Measurement flasks 100 ml
Conductance measurement cell (Fig. 3.4)
Conductance meter
Magnetic stirrer plate
Magnetic stirrer bar

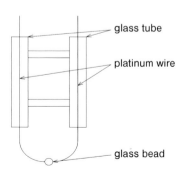
glass tube

platinum wire

glass bead

Fig. 3.4 Simplified conductance measurement cell.

Setup

Because no absolute conductivities but only relative changes are recorded a simplified conductance measurement cell as depicted may be employed.

Procedure

1. Titration curves of the following samples are recorded:

Sample (5 ml each, 1 M)	Titration solution (1 M)
HCl	KOH
CH$_3$COOH	KOH
HCl + CH$_3$COOH	KOH
CH$_3$COONa	HCl

Add 5 ml of sample into beaker, and add water until the platinum wires are completely immersed even when stirrer is operating.

Switch on conductance meter; select range where a conductance close to the upper limit of the range is displayed; do not change range during titration (for simplicity and to avoid recalculations).

Add titrant in 0.5-ml increments.

2. Determine the amount of HCl and CH$_3$COOH in samples of unknown composition.

Add water to the sample in the measurement flask up to mark.

Take 20 ml of this solution and titrate with 1 **M** KOH solution.

Evaluation

Plot titration curves (Fig. 3.5) and determine equivalence points.

Calculate content of HCl and CH$_3$COOH in samples of unknown composition.

At $V_{KOH} = 4.14$ ml and 9.5 ml turning points are observed. The mass of hydrochloric acid is

$$m_{HCl} = 5 \cdot V_{KOH} \cdot c_{KOH} \cdot M_{HCl} = 5 \cdot 4.14 \cdot 10^{-3} \cdot 36.458 = 0.754 \text{ g}$$

The original mass was 0.73 g. The calculation for the mass of acetic acid is

$$m_{HAc} = 5 \cdot V_{KOH} \cdot c_{KOH} \cdot M_{HAc} = 5 \cdot 5.36 \cdot 10^{-3} \cdot 60.1 = 1.61 \text{ g}$$

Initially 1.5 g acetic acid were present in the sample.

Questions

- Explain the expected titration curves taking into account the titration reactions and the ionic conductivities.
- Is conductometric indication a viable method for precipitation titration (e.g., BaCl$_2$ with K$_2$SO$_4$, KCl with AgNO$_3$)? Describe the expected shape of titration curve (compare Expt. 4.4).

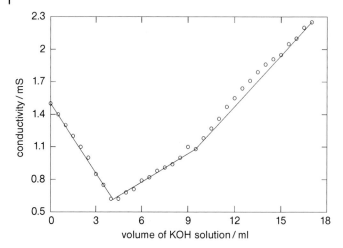

Fig. 3.5 Typical titration curve of a solution with unknown concentration of acetic and hydrochloric acid and a solution of 1 **M** KOH as titrant.

- Is conductometric titration a suitable method for titration of weak acids with weak bases (or vice versa)? Why?

Literature

P. W. Atkins, J. de Paula, Physical Chemistry, 8th ed., Oxford University Press, Oxford 2006, p. 1019.

M. R. Wright: An Introduction to Aqueous Electrolyte Solutions, John Wiley & Sons, Chichester 2007.

Experiment 3.5: Chemical Constitution and Electrolytic Conductance

Task

Constitution isomerism of an aliphatic nitro compound is studied by conductometry.

Fundamentals

According to its constitutional formula nitroethane ($CH_3–CH_2NO_2$) does not have an acidic proton. Its aqueous solution shows only a very small electrolytic conductance. Upon addition of sodium hydroxide nitroethane behaves like an acid, and the respective sodium salt is formed according to

$$CH_3–CH_2NO_2 \rightarrow CH_3–CH=NOOH \tag{3.15}$$

$$CH_3–CH=NOOH + NaOH \rightarrow CH_3–CH=NOONa + H_2O \tag{3.16}$$

The second reaction [Eq. (3.16)] is a neutralization reaction, accordingly it is very fast. The rate of the first reaction remains unknown. Measurement of the time-dependent electrolytic resistance or conductance with a simple setup (see

Expt. 3.4) enables an estimate to be made. A slowly growing resistance indicates the slow isomerization reaction, followed by the fast neutralization reaction associated with consumption of NaOH and an increase in cell resistance.

The following reactions are possible when hydrochloric acid is added to the reaction mixture:

$$CH_3-CH=NOONa + HCl \rightarrow NaCl + CH_3-CH=NOOH \tag{3.17}$$

$$CH_3-CH=NOOH \rightarrow CH_3-CH_2NO_2 \tag{3.18}$$

Again, time-dependent measurements of the cell resistance may provide information about the relative rates of both reactions. Because the reaction of the sodium salt with HCl is an ionic one it is presumably fast, quickly yielding a constant concentration of NaCl and the corresponding cell resistance. The following isomerization is much slower, associated with a slower growth of cell resistance.

Execution
Chemicals and instruments
10 ml of an aqueous solution of nitroethane 0.1 **M**
10 ml of an aqueous solution of NaOH 0.1 **M**
10 ml of an aqueous solution of HCl 0.1 **M**
Conductance measurement cell
Conductance meter
Cryostat or other device for cooling the reaction mixture

Setup
See Expt. 3.4.

Procedure
10 ml of an aqueous solution of nitroethane 0.1 **M** and 10 ml of an aqueous solution of NaOH 0.1 **M** are mixed at a temperature of 0 °C. The cell resistance (or conductance) is measured at one-minute intervals. When a constant value is reached the mixture is warmed to 25 °C, 10 ml of an aqueous solution of HCl 0.1 **M** are added. Again the cell resistance is measured at one-minute intervals. For comparison finally the cell resistance is measured of a mixture of 10 ml of an aqueous solution of NaOH 0.1 **M**, 10 ml of an aqueous solution of HCl 0.1 **M** and 10 ml of water.

Evaluation
A typical set of results is displayed in Fig. 3.6. The plot obtained during the first reaction according to Eqs. 3.15 and 3.16 indicates a relatively slow isomerization reaction. A corresponding situation is observed during the second reaction. Taking into account the considerably higher temperature (25 °C instead of 0 °C) in the second reaction the forward reaction in the equilibrium

$$CH_3-CH_2NO_2 \leftrightarrows CH_3-CH=NOOH \tag{3.19}$$

is much faster.

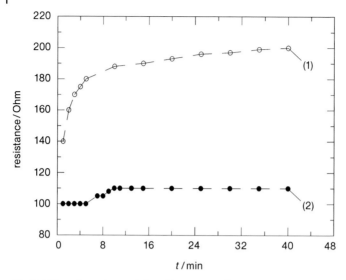

Fig. 3.6 Time dependence of the cell resistance during reaction of nitroethane with sodium hydroxide (1) and during the subsequent reaction with HCl (2).

The assumption of a complete conversion of sodium nitroethane into the nitro form of nitroethane is supported by the agreement of the resistance of the reaction mixture observed when reaching a constant value (see trace (2) in Fig. 3.6) with the resistance $R = 107\ \Omega$ of the mixture of 10 ml of an aqueous solution of NaOH 0.1 **M**, 10 ml of an aqueous solution of HCl 0.1 **M** and 10 ml of water.

Experiment 3.6: Faraday's Law

Task
Measurement of electrolytic conversions at electrodes

Fundamentals
As suggested by W. Nernst, the combination of an electron-conducting and an ion-conducting phase (e.g., a copper rod and a copper sulfate solution) is called an electrode. The combination of two electrodes forms an electrochemical cell. When an electric current is flowing through the cell at these electrodes (i.e. at the phase boundaries) chemical reactions must proceed interconnecting ion and electron conduction. A flow of electrons from the electrode into the external electric circuit results from the release of electrons from electrolyte solution constituents or from the electrode material (e.g., a metal) itself. Accordingly an

oxidation proceeds: this electrode is called anode. This process may be supported by anions supplied from the solution.

At the second electrode electrons are transferred from the electrode into the electrolyte solution, i.e. to one of its constituents. This process is a reduction: the electrode is called cathode. A typical example is the reduction of metal cations, and these may again be supplied from the electrolyte solution.

A detailed analysis of mass changes proceeding during electrolysis reveals that mass changes are proportional to the transferred electric charge. This was observed for the first time by M. Faraday, and is known as the First Faraday Law. Several electrolysis cells may be connected in series, and a sufficiently large electric voltage is now needed to keep electrolysis running. The same amount of charge is transferred through all cells, and according to the composition of the cells different processes may be observed. When comparing, e.g., the mass m of cathodically deposited hydrogen, copper, and silver a ratio $m_H : m_{Cu} : m_{Ag} = 1 : 31.8 : 107.9$ is observed. This is equivalent to the ratio of molar masses divided by the ionic charge number z: $(2/2) : (53.6/2) : (107.9/1)$; these quotients were formerly called equivalent weights. The relationship is stated in Faraday's Second Law: The masses of different substances electrochemically formed from different electrolytes is proportional to the respective molar mass divided by the ionic charge number (also called molar mass of an ion-equivalent).

The consumed charge Q depends on the time of electrolysis and the current

$$Q = t \cdot I \tag{3.20}$$

When m grams of ions are discharged this is equivalent to $n = m/M$ mol of these ions with molar mass M; with an ionic charge number z $(m/M) \cdot z$ moles of electrons or $(m/M) \cdot z \cdot N_A$ electrons are consumed. One mol of electrons with the elementary charge q_e (sometimes also called e_0) amounts to

$$N_A \cdot q_e = N_A \cdot e_0 = 96484 \ A \cdot s = 96484 \ C \tag{3.21}$$

This amount of charge is called 1 Farad (F). Faraday's first law can thus be rewritten

$$m = I \cdot t \, (M/(z \cdot F)) \tag{3.22}$$

and the second law

$$m_1/m_2 = (M_1/z_1)/(M_2/z_2) \tag{3.23}$$

These relationships were employed in the early years of electricity distribution to measure the charge passed through an electric circuit; applied in a coulometer they were the precursors of today's electric meter. Initially called Stia counters they are still in use for monitoring hours of operation of devices.

Execution

Chemicals and instruments

Aqueous solution of K_2SO_4 (10 wt%)

Aqueous solution of $CuSO_4$ 1 **M**

Aqueous solution of $AgNO_3$ 1 **M**

Constant current supply 100 mA

Multimeter

Oxyhydrogen coulometer [3]

2 Copper electrodes

Silver electrode

Platinum electrode

Setup

The following figure Fig. 3.7, shows the setup schematically. Instead of the oxy-hydrogen coulometer another device for water electrolysis could be employed provided that precise determination of the amount of developed gas is possible.

Procedure

- Determine the mass of the metal electrodes; adjust level in eudiometer to zero.
- Connect cells according to scheme, pass a current of 100 mA for at least 20 min.

Measure the amount of oxyhydrogen gas; remove electrodes from cells, rinse them carefully, dry, and determine mass.

Evaluation

Calculate the converted (deposited) masses according to Faraday's Law and compare the values with your own results.

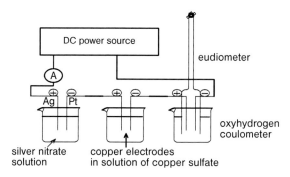

silver nitrate solution

copper electrodes in solution of copper sulfate

Fig. 3.7 Setup for the examination of Faraday's Laws.

---

3) Sometimes also called combustion-gas coulometer.

Experiment 3.7: Kinetics of Ester Saponification

Task

The rate of reaction, the activation energy, and the preexponential factor in the Arrhenius equation of a chemical reaction are determined by conductance measurements.

Fundamentals

The experimental determination of the rate constant k of the alkaline ester saponification of ethyl acetate and its dependence on temperature starts with the reaction equation

$$CH_3CO_2C_2H_5 + K^+ + OH^- \xrightarrow{k} CH_3CO_2^- + K^+ + C_2H_5OH \qquad (3.24)$$

In this experiment we start with an equimolar mixture of ethyl acetate and potassium hydroxide. During the reaction hydroxyl ions are consumed and acetate ions are generated, whereas the concentration of potassium ions stays constant. Because the former ions have significantly different equivalent conductivities, progress of the reaction can be monitored by conductance measurements of the reaction mixture. From measurements of $\kappa = \kappa(t)$ the rate constant of the reaction k can be calculated.

Evaluation of data is based on Eq. 3.42 (see below) and a plot of $1/(\kappa_0 - \kappa(t))$ against $1/t$. The slope of the graph yields k. Determination of k at different temperatures enables an Arrhenius plot to be drawn, and this in turn provides the energy of activation and the preexponential factor of the Arrhenius equation.

The specific conductance $\kappa = \kappa(t)$ at a time t is given with a value of κ_0 at $t=0$ by

$$\kappa = \kappa_0 - \text{contribution of consumed ionic concentration}$$
$$+ \text{contribution of generated ionic concentration} \qquad (3.25)$$

The contribution of ions to the specific conductance is given by

$$\Lambda_{eq} = \lambda_+ + \lambda_- = \kappa/(z \cdot c) \qquad (3.26)$$

with $z=1$, c in $mol \cdot cm^{-3}$ and

$$\lambda_{OH^-} \cdot c_{OH^-} \qquad (3.27)$$

and

$$\lambda_{Ac^-} \cdot c_{Ac^-} \qquad (3.28)$$

The concentration of hydroxyl ions consumed up to time t is (according to the reaction equation, equal to the concentration of generated acetate ions. It can be given by

$$\Delta c = c_0 - c \qquad (3.29)$$

With starting concentration c_0 and concentration c in $mol \cdot l^{-1}$ at time t. Eq. 3.26–3.29 can thus be rewritten

$$\kappa = \kappa_0 - \Delta c \cdot \lambda_{OH^-} \cdot 0.001 + \Delta c \cdot \lambda_{Ac^-} \cdot 0.001 \tag{3.30}$$

or

$$\Delta c = \frac{\kappa_0 - \kappa}{\left(\lambda_{OH^-} - \lambda_{Ac^-}\right) \cdot 0.001} \tag{3.31}$$

The factor 0.001 results from the conversion from the usual concentration value in $mol \cdot l^{-1}$ into the units $mol \cdot cm^{-3}$ used for conductance data. We assume that equivalent conductivities stay constant because the overall concentration of ions does not change markedly during the reaction thus we can simplify

$$\left(\lambda_{OH^-} - \lambda_{Ac^-}\right) \cdot 0.001 = A \tag{3.32}$$

and we obtain with Eq. 3.30

$$\Delta c = \frac{\kappa_0 - \kappa}{A} \tag{3.33}$$

The alkaline ester saponification is a second-order reaction according to

$$A + B \rightarrow C + D \tag{3.34}$$

With A=ester, B=hydroxyl ions; C=acetate and D=ethanol, these symbols are used as labels; and with stoichiometric factors equal to unity the reaction rate is given by

$$v = -\frac{dc_A}{dt} = k \cdot c_A \cdot c_B \tag{3.35}$$

Because

$$c_A = c_B = c_{OH^-} = c \tag{3.36}$$

Eq. 3.35 can be simplified to

$$v = -\frac{dc}{dt} = k \cdot c^2 \tag{3.37}$$

or

$$-\frac{dc}{c^2} = k \cdot dt \tag{3.38}$$

Integration from $t=0$ to t yields

$$-\frac{1}{c} + \frac{1}{c_0} = -k \cdot t \tag{3.39}$$

The fraction Δc consumed at time t with a starting concentration c_0 is given by

$$\Delta c = c_0 - c \tag{3.29}$$

Eq. 3.39 can be inserted, and after rearrangement we obtain

$$\Delta c = c_0 \cdot \left(1 - \frac{1}{1 + c_0 \cdot k \cdot t}\right) \tag{3.40}$$

The reaction rate constant k can be obtained based on the relationship between k and κ. This can be done by combining Eqs. 3.33 and 3.40

$$\frac{\kappa_0 - \kappa}{A} = c_0 \cdot \left(1 - \frac{1}{1 + c_0 \cdot k \cdot t}\right) \tag{3.41}$$

After rearrangement and with $B = A \cdot c_0$ we obtain

$$\frac{1}{\kappa_0 - \kappa} = \frac{1}{c_0 \cdot k \cdot t \cdot B} + \frac{1}{B} \tag{3.42}$$

A plot of $1/(\kappa_0 - \kappa)$ vs. $1/t$ yields as abscissa the constant $1/B$. k can be calculated from the slope $1/c_0 \cdot k \cdot B$ of the plotted line.

Execution
Chemicals and instruments
Aqueous solution of ethyl acetate 0.125 **M**
Aqueous solution of KOH 0.125 **M**
Conductometer
Conductance measurement electrode
Thermostat
Magnetic stirrer
Magnetic stirrer bar
Measurement cell with water jacket
Stop watch
Pipette
Thermometer

Procedure
The cell constant of the conductance measurement cell is determined as described in Expt. 3.3. For determination of κ_0, 90 ml of ultrapure water are poured into the measurement cell, 10 ml of the solution of KOH are added. The solution is thoroughly mixed with the added magnetic stirrer bar; now κ_0 is determined. After emptying, cleaning and drying the cell carefully 80 ml of ultrapure water are added. 10 ml of the solution of KOH are added under vigorous stirring. When the temperature has settled at the desired value, 10 ml of the solution of ethyl acetate are added quickly. The stop watch is started, and the electrode is moved up and down a few times to enhance mixing of the solution. The stirrer is operated at high speed. After 1, 5, 10, 20, 30 and 60 min,

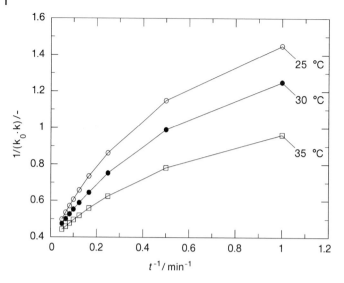

Fig. 3.8 Plot of conductance data obtained during alkaline ester saponification.

conductance values are recorded (note changes of the range of the conductometer). The procedure is repeated at $T=15\,°C$[4], $35\,°C$, and $50\,°C$ or other possible temperatures spaced as far apart as feasible.

Evaluation

A typical plot is shown in Fig. 3.8.

At $T=25\,°C$ a slope of 1.0 and an abscissa of 0.52 were determined; at $T=30\,°C$ the respective values were 0.82 and 0.48; at $T=35\,°C$ they were 0.5 and 0.44. The rate of reaction at $T=25\,°C$ is $k=41.6\ l\cdot mol^{-1}\cdot min^{-1}$; at $T=30\,°C$ $k=46.8\ l\cdot mol^{-1}\cdot min^{-1}$ and at $T=35\,°C$ $k=64\ l\cdot mol^{-1}\cdot min^{-1}$. A literature value $k=0.111\ l\cdot mol^{-1}\cdot s^{-1}$ is available. From the energy of activation E_a and the rate constant k at $T=35\,°C$ the value of $k_{70\,°C}$ should be determined. This experiment does not yield very precise values, and a statistical assessment is recommended.

Questions

Which values may be discarded when some data are deviating from a straight line?

Literature

A. J. Kirby in: Compr. Chem. Kin. Vol. 10 (C. H. Bamford and C. F. H. Tripper, eds.), Elsevier, Amsterdam 1972.

4) If no cryostat is available this value may be replaced by a slightly higher one. In order to maintain a wide temperature range the highest temperature should be increased if possible.

Experiment 3.8: Movement of Ions and Hittorf Transport Number

Tasks
- The transport number t and the equivalent conductance at infinite dilution λ_0^- of sulfate ions is determined in the electrolysis of an aqueous solution of 0.1 M H_2SO_4.
- The ionic mobility, the equivalent conductance, and the ionic radius of the permanganate ion are determined.

Fundamentals
Transport of electricity through an ionic conductor, e.g., an electrolyte solution, a molten salt, or a solid electrolyte, is associated with movement of ions. The accelerating force of the electric field E [5] is countered by the braking force of Stokes friction. At equilibrium for an ion with radius r_i in a medium of viscosity η a constant traveling velocity (drift speed) v is established:

$$v = \frac{z \cdot e_0 \cdot E}{6 \cdot \pi \cdot \eta \cdot r_i} \tag{3.43}$$

The drift speed standardized with respect to the strength of field is called mobility

$$u = v/E \tag{3.44}$$

With Eq. 3.44, data from experimental determination of traveling velocity obtained from, e.g., the speed of the movement of a boundary indicated by colored ions (compare Expt. 3.1) the ionic radius r_i can be obtained. From the relationship between ionic mobility and ionic equivalent conductance the latter can be obtained:

$$\lambda_{eq} = u \cdot F \tag{3.45}$$

If all transport properties of ions (both anions and cations) are equal as in the case of a 1:1 electrolyte the flow of one mole of electrons (i.e. 96485 Cb or 96485 A·s) charge would result in the movement of 0.5 mol of anions towards the anode and 0.5 mol of cations toward the cathode. In reality transport properties of ions are significantly different. The particularly high (extra) conductance of protons and hydroxyl ions is caused by a special mechanism of movement (Grotthus jump mechanism, P. W. Atkins, J. de Paula, Physical Chemistry, 8th ed., Oxford University Press, Oxford 2006, p. 766; see also: EC:34) enabling these ions to contribute more towards charge transport than other ions.

Experimental investigations of these details have resulted in transference numbers t, sometimes also called Hittorf transport numbers honoring its initial

---

[5] Because only the accelerating force of the electric field in the direction of movement is considered, a vectorial display of E is not necessary.

definer (P.W. Atkins, J. de Paula, Physical Chemistry, 8th ed., Oxford University Press, Oxford 2006, p. 768). The transference number t_+ designates the fraction I_k of the total current I according to $I_k = t_+ \cdot I$. Obviously the sum of transference numbers for a given electrolyte must be one:

$$t_+ + t_- = 1 \tag{3.46}$$

As first described by Hittorf the determination of t_+ can be done in a simple electrolysis experiment with carefully separated anode and cathode compartments (effected by a diaphragm or a special cell construction). Determination of concentrations of electrolyte constituents before and after electrolysis in both compartments and of the charge transported, the fraction of charge carried by anions and cations can be determined. Careful distinction between processes with and without consumption (i.e. chemical conversion by reduction or oxidation) of ions is required. An example of the former type is an experiment with hydrochloric acid, and electrolysis with sulfuric acid is an example of the latter kind. Whereas protons yield hydrogen at the cathode as expected sulfate ions will not be oxidized, and instead oxygen is formed. This decoupling must be taken into account in the evaluation of experimental results (EC:30).

In this experiment sulfuric acid is electrolyzed. The change of concentration of protons is evaluated. From a titration of the solutions in the cathode and anode compartment before and after electrolysis the number of moles n_A and n_C and their respective changes Δn can be obtained. From the average of the change of mole numbers

$$\bar{\Delta n} = (|\Delta n_A| + |\Delta n_K|)/2 \tag{3.47}$$

the charge transported by the anions q_-

$$q_- = \bar{\Delta n} \cdot F \tag{3.48}$$

can be calculated. Because the total charge q transported by anions and cations deduced from the duration of the electrolysis and the flowing (constant) current (or from the charge determined with a coulometer) the transference number can be calculated according to

$$t_- = q_-/(q_- + q_+) \tag{3.49}$$

According to

$$\lambda_{0,SO_4^{2-}} = t_- \cdot \lambda_{0,H_2SO_4} \tag{3.50}$$

the ionic equivalent conductance of the sulfate ion at infinite dilution can be calculated.

Execution

Chemicals and instruments

0.1 **M** aqueous solution of H_2SO_4

0.1 **M** aqueous solution of KOH

Aqueous solution of H_2SO_4 20wt.%

0.005 **M** aqueous solution of $KMnO_4$

0.005 **M** aqueous solution of KNO_3

Urea

Tools and chemicals for acid-base titration

Adjustable electrolysis cell with platinum electrodes; the plugs holding the electrode must permit passage of gas from the cell

DC power source 40 V

Oxyhydrogen or digital coulometer

First Task

Setup

The setup for the first task is depicted schematically in Fig. 3.9.

Procedure

Preparation of electrolysis cell

The lower part of the electrolysis cell mounted in the adjustable holder is filled with 0.1 **M** aqueous solution of H_2SO_4 up to a level slightly above the valves. After closing the valves liquid above the valves is removed carefully with a pipette.

Anode and cathode chamber are filled up to the marks with the same acid solution; conveniently, initially a fixed volume is added with a pipette, and further solution is added with a burette. Now the valves are opened again. If the solution levels are no longer at the marks, the vessel must be tilted slightly until the levels are adjusted to the marks.

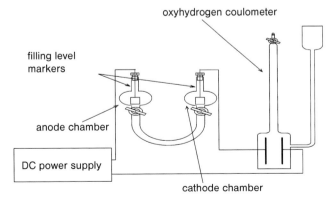

Fig. 3.9 Experimental setup for the determination of transport numbers.

Preparation of the oxyhydrogen coulometer
(This device is a rather old-fashioned approach towards charge measurements; today precise digital coulometers are available. In this experiment the coulometer serves as a control of the passed charge, which can be calculated easily of course by multiplying the adjusted current[6] with the passed time. In addition it is a nice exercise of the application of Faraday's laws.) The oxyhydrogen coulometer is filled with an aqueous solution of H_2SO_4 20wt%. With the valve at the burette open the upper meniscus is adjusted to the zero mark by moving the balancing vessel up or down. The valve is finally closed. The recorded gas volume at the end of the experiment must be converted to standard conditions using the ideal gas law and the actual temperature and ambient air pressure.

The power supply, the coulometer, and the electrolysis cell are connected. The voltage at the power supply is set to zero, the power supply is switched on, and the current is adjusted to 25 mA. A reasonable duration of the electrolysis is 120 min.

End of the experiment
After closing the valves the power supply is switched off. The electrodes are removed. Anode and cathode chamber are closed with plugs, the cell is removed from the holder and shaken carefully (why?). Samples are taken from both chambers and titrated. In addition a sample of the 0.1 **M** aqueous solution of H_2SO_4 is titrated. In the protocol, the volumes of both electrode chambers, the temperature, the passed charge (from time and current and from readings at the coulometer), and the final concentrations in the electrode chambers must be recorded. From the concentrations the masses are calculated as number of moles *n*.

Evaluation
The following data are needed for evaluation:
- Volume of electrode chambers
- Temperature
- Starting concentration of sulfuric acid
- Concentration of sulfuric acid in anode and cathode chamber after electrolysis
- Electrolysis current and duration of electrolysis
- Volume of formed oxyhydrogen gas.

From Eq. (3.47) the mean change of mole numbers in both electrode chambers is calculated. From Eq. (3.49) the charge transported by the anions can be calculated. After calculating the total charge passed (from $I \cdot t$ and from the volume of produced oxyhydrogen gas) the transport number of the anion can be calcu-

---

6) The rather large DC voltage allows operating the power source as a constant current source by inserting a suitable resistor, otherwise the current must be monitored carefully during the experiment.

lated from Eq. (3.49). From the equivalent conductance of sulfuric acid (see tabulated values) the ionic conductance of the sulfate ion can be calculated.

In a typical experiment the starting concentration of sulfuric acid was $c=0.0985$ **M**. Electrolysis was performed at $I=25$ mA for $t=2$ h; thus a charge of $q=180$ A·s was passed. From the volume $V=36$ ml of produced oxyhydrogen gas converted to standard conditions a charge of $q=186.7$ A·s was calculated, in good agreement with the charge derived above. Because adjustment of the constant current in the used setup was only of limited precision, further calculations are based on the charge derived from the coulometer. After electrolysis the acid concentration in the anode chamber has changed to $c=0.10265$ **M**; in the cathode chamber the final value was $c=0.09515$ **M**. Taking into account the liquid volume in the cathode chamber (54.9 ml) and the anode chamber (51.3 ml) the change of the number of moles of protons in the cathode chamber is $\Delta n_C=-3.97\cdot10^{-4}$ mol and in the anode chamber $\Delta n_A=4.28\cdot10^{-4}$ mol. The average change is $\Delta n=4.125\cdot10^{-4}$ mol. The transport number of the sulfate ion is $t_-=0.213$. The value of $\lambda_{0,SO_4^{2-}}^+=183.2$ cm$^2\cdot\Omega^{-1}\cdot$mol^{-1} is in good agreement with literature values.

Second Task

Setup
The setup for the second task is depicted schematically in Fig. 3.10. The distance between the electrodes was $d=32.5$ cm in a typical setup.

Procedure
The upper reservoir volume is filled with a 0.005 **M** aqueous solution of KMnO$_4$; in the setup used here 100 ml were needed. 3 g of urea were added in order to increase density. The valve is opened until the liquid fills the middle tube down to the glass bead. Aqueous solution of KNO$_3$ is filled into one of the ports of the U-shaped tube until half-filled without mixing with the solution of KMnO$_4$. (If excessive mixing occurs the procedure must be repeated.) The valve is opened carefully again, and the level of solution of KMnO$_4$ will slowly rise on both sides. The glass bead helps to avoid turbulence; thus well-defined, sharp

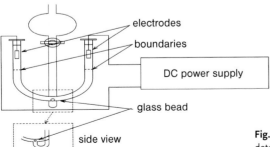

Fig. 3.10 Experimental setup for the determination of traveling velocity.

boundaries between the solutions are established. The position of the bound-aries is marked on the U-shaped tubes. After inserting the electrodes a DC-volt-age of about $U=40$ V is applied. The shift of the positions of the boundaries after 5, 10, 15, and 20 min is determined on both sides, and the average shift is calculated.

Evaluation

From the distance between the electrodes ($d=32.5$ cm) and the applied voltage, the strength of field E is calculated. From the traveling velocity of the perman-ganate ion (visible as movement of the phase boundary) the ionic mobility can be derived. The calculation of the equivalent conductance and the ionic radius follows Eq. (3.44) and (3.43). The viscosity of the solution is assumed to be equal to the viscosity of water at the temperature of the experiment (literature value is used).

In a typical experiment with the given distance between the electrodes, $U=40$ V is applied. The following average shifts of the phase boundary is ob-served:

Time/min	Shift/cm
5	0.55
10	0.85
15	1.1
20	1.4

The velocity of the shift v is obtained by averaging over the duration of the ex-periment; the mobility u is in good agreement with literature values

$$u = \frac{v}{U} = \frac{v \cdot d}{U} = \frac{0.76 \cdot 10^{-3} \cdot 32.5}{40} = 0.88 \cdot 10^{-3} \, cm^2 \cdot V^{-1} \cdot mol^{-1} \qquad (3.51)$$

The equivalent conductance is

$$\lambda_{eq} = u \cdot F = 0.76 \cdot 10^{-3} \cdot 96464 = 73.3 \, cm^2 \cdot \Omega^{-1} \cdot mol^{-1} \qquad (3.52)$$

The ionic radius is according to

$$r_i = \frac{z \cdot e_0 \cdot E \cdot t}{6 \cdot \pi \cdot \eta \cdot v} \qquad (3.53)$$

$r_i = 125$ pm. A comparison with ion radii determined by crystallography implies a low degree of solvation with the associated increase of the effective ion radius.

Questions
• How can you determine ionic conductance from the transport number and the equivalent conductance?
• Are there other ways to ionic conductivities?

- Is the choice of electrolyte in the present experiment important in the determination of the transport number? Is the evaluation different when HCl would have been selected?

Literature
M. R. Wright: An Introduction to Aqueous Electrolyte Solutions, John Wiley & Sons, Chichester 2007.

Experiment 3.9: Polarographic [7] Investigation of the Electroreduction of Formaldehyde

Task
The rate constant of the dehydration of formaldehyde hydrate is determined by polarography.

Fundamentals
Polarography can be employed in kinetic and mechanistic investigations provided that it involves in the reaction sequence electrochemically active species whose concentration may provide information about the ongoing process. Currents caused by their generation or consumption are called kinetic currents, and these are limited by the rate of a chemical reaction preceding or following the electrochemical charge transfer reaction. In the case of a preceding chemical reaction, electrochemically inactive species are converted into polarographically active ones, which can be reduced or oxidized at the mercury electrode. If this reaction is slower than the charge transfer reaction the kinetically limited current is controlled by the rate constant k of the chemical reaction.

The cathodic reduction of formaldehyde is a typical example. In aqueous solution this substance is present almost completely in its hydrated form as methylene glycol. Only free formaldehyde formed according to

$$CH_2(OH)_2 \underset{k_b}{\overset{k_f}{\rightleftharpoons}} CH_2O + H_2O \tag{3.54}$$

in an equilibrium with the hydrated form can be reduced at a mercury electrode. Its concentration [8] is given by

$$K = \frac{[CH_2O]}{[CH_2(OH)_2]} = \frac{k_f}{k_b} \tag{3.55}$$

Reduction proceeds according to

---

7) Polarographic methods are mostly employed in analytical applications, they are treated in detail in Chapter 4. This experiment is less analytical, thus it is presented

here. Further details of polarography are treated in Chapter 4.
8) Following concentrations are indicated with square brackets [].

$$2 \, H_2O + CH_2O + 2 \, e^- \rightarrow CH_3OH + 2 \, OH^- \tag{3.56}$$

The reaction is subject to general acid-base catalysis. Besides hydroxyl ions also Bronsted bases are catalytically active. The general rate equation is accordingly:

$$k_f = k_0 + k_H \cdot [H^+] + k_{OH} \cdot [OH^-] + \sum k_A \cdot [A^-] + \sum k_B \cdot [B^+] \tag{3.57}$$

k_f is the rate constant for the dehydration in a given solution composition; k_0 is the rate constant in neutral solution in the absence of catalytically effective ions; $k_H \cdot [H^+]$ describes the contribution of protons; $k_{OH} \cdot [OH^-]$ the contribution of hydroxyl ions; $\sum k_A \cdot [A^-]$ and $\sum k_B \cdot [B^+]$ describes the influence of further acidic or basic species.

Because the hydroxyl ions formed according to Eq. (3.56) might act autocatalytically, a buffered solution is used. Data of k_0 pertaining to a neutral, unbuffered solution are obtained by extrapolating data obtained at different buffer concentrations to zero buffer concentration.

A reaction layer thickness δ_r similar to the Nernstian diffusion layer thickness wherein the rate-determining reaction step (dehydration) proceeds is assumed for the calculation. The concentration of the hydrated form $CH_2(OH)_2$ within the reaction layer is constant and equal to c_0 when the rate of dehydration

$$\frac{d[CH_2O]}{dt} = k_f \left[CH_2(OH)_2 \right] \tag{3.58}$$

is small compared to the rate of diffusion of $CH_2(OH)_2$ into the reaction layer.

Assuming chemical equilibrium outside the reaction layer and a stationary state inside the layer a mean kinetically limited current at the dropping mercury electrode can be calculated

$$\bar{I}_k = 5.1 \cdot 10^{-3} \cdot n \cdot F \cdot \left[CH_2(OH)_2 \right] \cdot (m \cdot \tau)^{2/3} \cdot (D_{CH_2O} \cdot k_f \cdot K)^{1/2} \tag{3.59}$$

with m: flow rate of mercury in $mg \cdot s^{-1}$ and τ: drop time in s. In order to obtain $k_f \cdot K$ the current described by the Ilkovič equation obtained under the same experimental conditions assuming a conversion fast in comparison with the diffusion of the hydrate is calculated. In this case the current is obviously limited by diffusion:

$$\bar{I}_{lim,diff} = 607 \cdot n \cdot (D_{CH_2(OH)_2})^{1/2} \cdot m^{2/3} \cdot [CH_2(OH)_2] \cdot \tau^{1/6} \tag{3.60}$$

Assuming $D_{CH_2O} = D_{CH_2(OH)_2}$ the ratio of both currents yields

$$\frac{\bar{I}_k}{\bar{I}_{lim,diff}} = 0.81 \cdot (\tau \cdot K \cdot k_f)^{1/2} \tag{3.61}$$

With the known equilibrium constant K the rate constant k_f can be calculated.

Rate (kinetic) control of the polarographically determined current can be demonstrated easily by measuring the current at different heights of the mercury supply vessel, i.e. at different values of τ. Rearrangement of the Ilkovič equation [Eq. (3.60)] yields

$$\bar{I}_{\text{lim,diff}} = 607 \cdot n \cdot (D_{\text{CH}_2(\text{OH})_2})^{1/2} \cdot [\text{CH}_2(\text{OH})_2] \cdot (m \cdot \tau)^{1/6} m^{1/2} \tag{3.62}$$

$m \cdot \tau$ is equal to the mass of a drop of mercury.

This depends only on the properties of the capillary inner diameter, surface tension, etc., but not on the height of the mercury vessel or the drop time. Thus

$$\bar{I}_{\text{lim,diff}} \approx m^{1/2} \tag{3.63}$$

A similar proportionality (dependence on height or drop time) cannot be observed with the kinetic current (compare Eq. 3.59).

Execution
Chemicals and instruments
Aqueous stock solution of 0.025 M NaH_2PO_4
Aqueous stock solution of 0.025 M Na_2HPO_4
Aqueous solution of formaldehyde 36%
Polarograph for DC polarography
Dropping mercury electrode
X-Y-recorded
Pipette 10 ml
Burette 2 ml
8 measurement flasks 100 ml

Setup
A 3-electrode arrangement (see Fig. 4.19) is used.

Procedure [9]
1. Measurement of polarograms of a solution $6.2 \cdot 10^{-2}$ M in CH_2O in phosphate buffer with c_{buff} (variation of buffer concentration needed for extrapolation to buffer concentration zero); method: DC polarography, filter 1 s; $E_{\text{Ag/AgCl}}$ = -1–1.8 V, range 5 µA; 10 mV $\cdot$ s^{-1}; $\Delta U = 50$ mV.

a) $c_{\text{buff}} = 0.0025$ M (10 ml each of both buffer stock solution + 0.5 ml of formaldehyde solution filled up to 100 ml).
b) $c_{\text{buff}} = 0.005$ M (20 ml each of both buffer stock solution + 0.5 ml of formaldehyde solution filled up to 100 ml).
c) $c_{\text{buff}} = 0.0075$ M (30 ml each of both buffer stock solution + 0.5 ml of formaldehyde solution filled up to 100 ml).

9) The following description pertains to a particular polarograph and cell volume. Quantities must be adapted if necessary, a simple mercury dropping electrode is sufficient for this experiment.

2. Measurement of the influence of the drop time on the kinetically limited current; solutions as in c)

a) $\tau=1$ s
b) $\tau=0.5$ s
c) $\tau=0.2$ s

Evaluation

The kinetically limited currents of formaldehyde reduction are plotted as a function of buffer concentration, and are extrapolated to zero buffer concentration.

Using Eq. (3.60) the diffusion-limited current theoretically obtained in the case of an infinitely fast preceding (dehydration) reaction is calculated. With the polarograph employed here, mercury flow rates m were found as a function of drop time τ:

τ/s	$m/mg\cdot s^{-1}$
1	3
0.5	5.3
0.2	12.8

(These values can also be determined approximately by measuring the mass of mercury dropping within a given time from the capillary into the bottle through air.)

The diffusion coefficient D of $CH_2(OH)_2$ is assumed as being equal to the value determined for methanol: $D=1.6\cdot10^{-5}$ $cm^2\cdot s^{-1}$. Using Eq. (3.55) the rate constant k_f of the dehydration of formaldehyde hydrate is calculated. The equilibrium constant K can be taken from the literature (P. Valenta, Collection Czech. Chem. Commun. **25** (1960) 853) $K=4.4\cdot10^{-4}$. Typical polarograms obtained at various concentrations of buffer are shown below (Fig. 3.11):

A plot of the kinetic current $\bar{I}_k$ as a function of buffer concentration is shown in Fig. 3.12.

The value of $\bar{I}_k$ at zero buffer concentration obtained by extrapolation is $\bar{I}_k=0.605$ μA. From the characteristic data of the dropping mercury electrode a value of $\bar{I}_{lim,diff}=805$ μA can be calculated. This yields a rate constant of $k_f=3.9\cdot10^{-3}$ s^{-1} in good agreement with the literature value $k_f=3.4\cdot10^{-3}$ s^{-1}.

The measurement of the limiting current at different drop times does not show any dependence on the drop time as expected for a kinetically limited current as shown in Fig. 3.13.

Literature

J.E. Crooks and R.S. Bulmer, J. Chem. Educ. **45** (1968) 725.
R. Brdicka, Collection Czech. Chem. Commun. **20** (1955) 387.
N. Landqvist, Acta Chem. Scand. **9** (1955) 867.
K. Vesely and R. Brdicka, Collection Czechoslov. Chem. Commun. **12** (1947) 313.

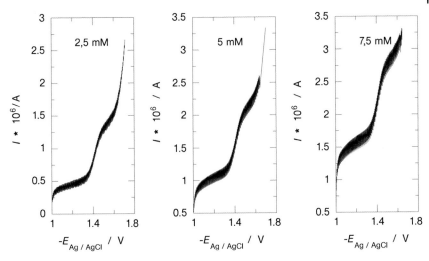

Fig. 3.11 Typical polarograms at different buffer concentrations.

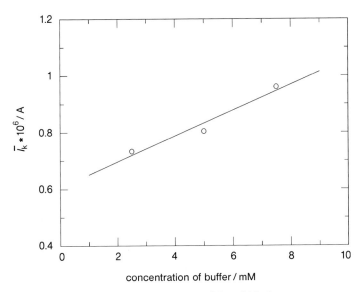

Fig. 3.12 Plot of kinetically limited currents of formaldehyde reduction as a function of buffer concentration.

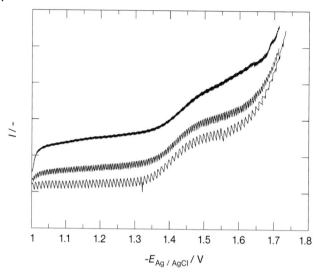

Fig. 3.13 Typical polarograms at different drop times: $\tau = 0.2$ s; $\tau = 0.5$ s; $\tau = 1$ s; from bottom to top, traces are offset for clarity.

Experiment 3.10: Galvanostatic Measurement of Stationary Current-Potential Curves

Task

The current density vs. electrode potential curves of dioxygen and hydrogen evolution at platinized platinum electrodes are measured and evaluated towards the determination of exchange current densities j_0.

Fundamentals

The relation between the anodic and cathodic partial current densities at an electrode and the current j_{ct} measured in an external electric circuit with the charge transfer overpotential η_{ct} is given by the Butler-Volmer equation:

$$j_{ct} = j_{ct,ox} - j_{ct,red} = j_0 \left\{ \exp \frac{a \cdot n \cdot F}{R \cdot T} \eta_{ct} - \exp \frac{(1-a)nF}{R \cdot T} \eta_{ct} \right\} \tag{3.64}$$

At an overpotential $\eta > R \cdot T / n \cdot F$ the counter reaction and accordingly the respective partial current density can be neglected. At a sufficiently large cathodic overpotential the equation is simplified:

$$j_{ct} = -j_0 \exp \frac{-(1-a)n \cdot F}{R \cdot T} \eta_{ct} \tag{3.65}$$

Logarithmic form and rearrangement yield

$$\eta_{ct} = \frac{RT}{(1-a)n \cdot F} 2.303 \lg j_0 - \frac{RT}{(1-a)n \cdot F} 2.303 \lg |j_{ct}| \qquad (3.66)$$

This equation resembles a general equation of the form

$$\eta_{ct} = A - B \lg |j_{ct}| \qquad (3.67)$$

According to its author it is called the Tafel equation, and the slope B is called Tafel slope. Without the rearrangement the equation has the form

$$\lg |j_{ct}| = \lg j_0 + \frac{(1-a) \cdot n \cdot F}{2.303 \cdot R \cdot T} |\eta_{ct}| \qquad (3.68)$$

A semi-logarithmic plot of $\lg |j_{ct}|$ vs $|\eta_{ct}|$ will yield j_0 from the abscissa and n as well as a from the slope. This approximation is limited at small currents by the counter reaction being no longer negligible; at high currents transport limitation (diffusion) becomes dominant over charge transfer limitation. In the range in between a graphical evaluation provides kinetic data of the electrode reaction.

In this experiment the respective diagrams of the oxygen and hydrogen evolution reaction at a platinum electrode are obtained and evaluated.

Execution
Chemicals and instruments
Aqueous solution of sulfuric acid **1 M**
Nitrogen purge gas
Adjustable current source (galvanostat)
High input impedance voltmeter
Platinized platinum electrode
Platinum electrode
Hydrogen reference electrode
H-cell

Setup
The platinized platinum electrode is mounted as a working electrode in the H-cell (see Fig. 1.4), and the platinum electrode serves as counter electrode. The hydrogen electrode is charged with hydrogen gas (see Fig. 1.3 and description) and inserted as reference electrode. The cell is filled with sulfuric acid solution. The current source is connected with the working and counter electrodes, and the voltmeter is connected with working and reference electrodes.

Procedure
The electrolyte solution is purged with nitrogen. Starting at $j=0$ mA the potential of the working electrode is measured. The current is increased stepwise to more negative values; at least three readings are taken per decade of current values. Before repeating the experiment with positive currents (oxygen evolution

reaction) the working electrode is set for three minutes to an electrode potential resulting in vigorous oxygen evolution.

Evaluation

Figure 3.14 shows a Tafel plot of the hydrogen evolution reaction.

The abscissa yields an exchange current density of $j_0 = 9.5 \ \mu A \cdot cm^{-2}$. This value is higher than that for smooth platinum as electrode material. From the slope, a Tafel slope of 14 mV per decade can be calculated. This value is in the range discussed by Vetter (K. J. Vetter, Angew. Chem. **73** (1961) 277) of approx. 30 mV per decade taken as an indicator of a reaction overpotential. This implies that the current density vs. electrode potential curve in the investigated range of electrode potentials is controlled by the recombination of adsorbed hydrogen atoms as the rate-determining step (rde) whereas the charge transfer reactions proceeds much faster. At smooth platinum a slope of 120 mV per decade is expected as typical for a simple charge transfer controlled reaction.

Figure 3.15 shows a Tafel plot of the oxygen evolution reaction.

Because the determination of the rest potential E_0 is difficult for the oxygen electrode the thermodynamically determined value of $E_0 = 1.229$ V was assumed. From the abscissa an exchange current density $j_0 = 1.3 \cdot 10^{-7} \ A \cdot cm^{-2}$ was calculated. The Tafel slope of 109 mV is close to the literature value of 120 mV per decade.

Literature

K. J. Vetter, Angew. Chem. **73** (1961) 277.
K. J. Vetter: Elektrochemische Kinetik, Springer, Berlin 1961.

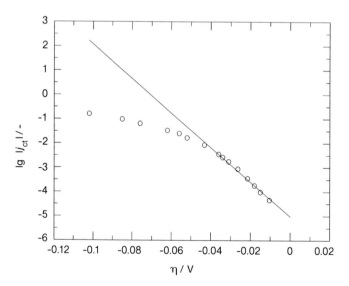

Fig. 3.14 Tafel plot of the hydrogen evolution reaction at a platinized platinum electrode in an aqueous solution of 1 **M** sulfuric acid.

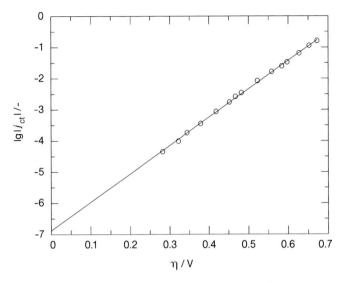

Fig. 3.15 Tafel plot of the oxygen evolution reaction at a platinized platinum electrode in an aqueous solution of 1 **M** sulfuric acid.

Experiment 3.11: Cyclic Voltammetry

Tasks
- Typical adsorption/desorption processes at a platinum electrode in contact
  with an aqueous solution of sulfuric acid are identified with cyclic voltammetry.
- The oxidation of organic molecules is studied with this method.
- The corrosion of nickel[10] in an aqueous electrolyte solution is monitored
  with cyclic voltammetry; breakthrough, Flade, and passivation potentials are
  measured.

Fundamentals
Cyclic voltammetry CV is an almost classical method in electrochemistry; it has
been established as a standard procedure for characterization of electrochemical
processes at the phase boundary electrode/electrolyte solution.

As already suggested by the name, the electrode potential E is swept cyclically
at a constant scan rate dE/dt between two limits (Fig. 3.16). The input voltage
(i.e. desired potential of the working electrode with respect to the reference electrode) applied to the system by a device called a potentiostat has the shape of a

---

10) Further experiments dealing with fundamental and applied
   aspects of corrosion can be found at the end of this chapter.

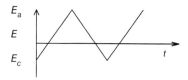

Fig. 3.16 Potential-time plot of the electrode potential in cyclic voltammetry.

triangle; thus the method has also been called the triangular voltage sweep method.

The potential limits are frequently the values at which the decomposition of the electrolyte solution (solvent or electrolyte) start; in he case of aqueous solutions it is the onset of hydrogen and oxygen evolution.

The method is a non-stationary one. In order to obtain the desired changes of electrode potential the concentrations of electrochemically active species involved in the establishment of the electrode potential at the electrochemical interface have to be fixed according to the Nernst equation. This is accomplished by a flow of electrical current across the working electrode under investigation and a third electrode called the counter or auxiliary electrode. Once the concentrations at the interface have reached values resulting in an electrode potential measured versus the reference electrode in agreement with the desired input value, the current drops to the value zero. The use of three electrodes: a working electrode (WE), a reference electrode (RE), and a counter electrode (CE) has resulted in the name three-electrode arrangement. A plot of the actual electrode potential versus the reference electrode (a plot of the input value may also be used, because assuming ideal regulation properties of the potentiostat this input value should be equal to the actual value) against the current yields a cyclic voltammogram. This plot yields electrode potential values indicative of the onset, cessation etc. of electrode processes like, e.g., oxidation, reduction of species from solution, metal dissolution, adsorption, desorption. Typical experiments involve investigations of redox systems, metal deposition or changes of coverage of the electrode with adsorbed species. From cyclic voltammograms (CVs) information about the thermodynamics of redox systems, the energetic positions of HOMOs and LUMOs, the kinetics of chemical reactions coupled to the electron transfer step and the rate of this electron transfer itself can be obtained.

In this experiment some basic qualitative studies are performed; in their interpretation creative interpretation of the observed data on the basis of literature data and of one's own considerations is required; less attention is paid to quantitative explanation.

The desired potential is obtained with the following experimental setup (see Fig. 3.17).

In this experiment, first CVs of a platinum electrode in contact with an aqueous solution of sulfuric acid will be recorded and interpreted. Next, the changes caused by the addition of an electrochemically active substance (formic acid) are explored. Finally CVs with a nickel working electrode are studied.

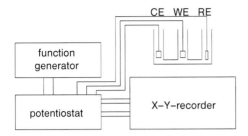

Fig. 3.17 Experimental setup for potentiostatic recording of CVs; CE: counter electrode; WE: working electrode; RE: reference electrode.

Execution
Chemicals and instruments
Aqueous solution of sulfuric acid 0.05 **M**
Aqueous solution of sulfuric acid 0.05 **M** + formic acid 0.1 **M** nitrogen purge gas
Potentiostat
Triangular voltage sweep generator
X-Y-recorder (as an alternative: PC with interface card)
2 platinum electrodes
Nickel electrode
Hydrogen reference electrode (HRE)
Three-electrode cell (H-cell)

Setup
With respect to the numerous conceivable combinations of instruments (potentiostats, function generators, recorders, PCs) a scheme of the wiring of instruments does not seem to be necessary.

Procedure
1. Recording a CV with supporting electrolyte solution only
The cell is filled with the aqueous solution of sulfuric acid, and platinum electrodes are inserted as working and counter electrodes; the hydrogen electrode (e.g., an electrode according to Will, see Fig. 1.3 p. 6) is used as a reference. Dissolved oxygen is purged from the electrolyte solution by bubbling nitrogen (or argon) through the gas inlet at the bottom of the working electrode compartment. The measurement itself is performed with a static solution (purge discontinued).

The current range at the potentiostat and the sensitivity at the recorder[11] are set to values resulting in the display of the complete CV in a single plot. The current flowing through the working and counter electrode causes a voltage drop at the shunt resistor used for current measurements; this voltage must be

---

11) Settings at the various instruments and components employed may differ; in the present description a rather generic procedure is given.

within the range of the recording device. The potential should be scanned in the range $0.02 < E_{RHE} < 1.66$ V. Before recording the CVs at a low scan rate for further discussion, the electrode surface is first brought into a reproducible state by applying a few fast potential scans ($dE/dt=1$ V·s^{-1}). A typical result is displayed below in Fig. 3.18.

A CV in the range $0.02 < E_{RHE} < 0.8$ V is recorded with enhanced current sensitivity. Finally CVs at different scan rates ($dE/dt=0.02 \ldots 0.1$ V·s^{-1} in 20 mV steps) are recorded in the range $0.3 < E_{RHE} < 0.5$ V.

Because there are no Faradaic reactions in this range in the investigated system, only charging of the double-layer capacity of the electrode/solution interface proceeds, ad accordingly this potential range is called the double-layer region. The CV at the highest scan rate should be recorded first because it results in the highest current values. Before every new CV plot a few scans should be run. A typical result is displayed above (Fig. 3.19).

2. CV of formic acid
The H-cell is filled with a aqueous solution of sulfuric acid 0.05 **M** + formic acid 0.1 **M**; all electrodes are used as before. A typical result is shown in Fig. 3.20.

3. CV of a nickel electrode in contact with an aqueous electrolyte solution of 0.05 **M** H$_2$SO$_4$

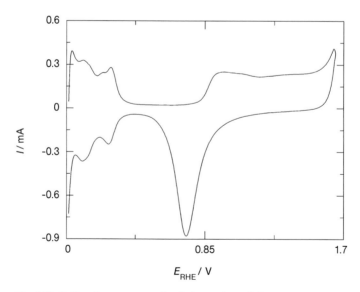

Fig. 3.18 Cyclic voltammogram of a platinum electrode in contact with an aqueous electrolyte solution of 0.05 **M** H$_2$SO$_4$, $dE/dt=0.1$ V·s^{-1}, nitrogen purged.

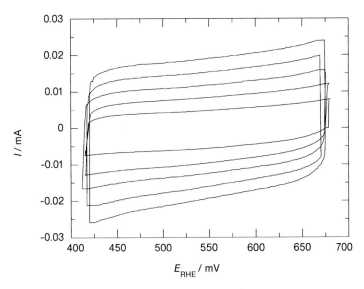

Fig. 3.19 CVs of a platinum electrode in contact with an aqueous electrolyte solution of 0.05 **M** H_2SO_4 in the double-layer region, $dE/dt = 0.02 \ldots 0.1$ V·s^{-1} from inward out, nitrogen purged.

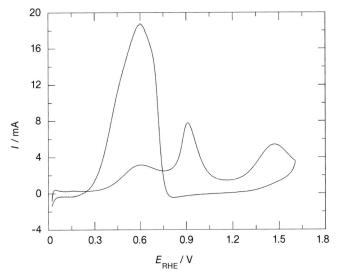

Fig. 3.20 CVs of a platinum electrode in contact with an aqueous electrolyte solution of 0.05 **M** H_2SO_4 + 0.1 **M** formic acid, $dE/dt = 0.1$ V·s^{-1}, nitrogen purged.

At a scan rate $dE/dt = 0.02 \ldots 0.1$ V·s^{-1}, the CV of a nickel electrode (nickel wire, spatula made of nickel) in contact with the electrolyte solution of 0.05 **M** H_2SO_4

is recorded. Starting at the negative potential limit three subsequent potential scans are recorded.

Evaluation

1. CV of the platinum electrode in supporting electrolyte solution

Potential range $0.02 < E_{RHE} < 1.66$ V
The potential has been scanned between the onset of hydrogen evlution and the onset oxygen evolution. In the positive-going scan starting at the negative limit, initially the hydrogen atom coverage of the electrode is oxidized. Two distinctly separated current peaks indicate the presence of two differently adsorbed (different surface sites, different strength of adsorption) types of H_{ad}. In the range $0.3 < E_{RHE} < 0.8$ V only charging current is observed. At about $E_{RHE} = 0.8$ V the formation of a layer of chemisorbed oxygen starts. Around $E_{RHE} = 1.6$ V oxygen evolution starts. In the negative-going scan the oxygen coverage is reduced with a considerable overpotential of several hundred millivolts. Around $E_{RHE} = 0.35$ V formation of the layer of adsorbed atomic hydrogen starts.

Potential range $0.02 < E_{RHE} < 0.8$ V
The more detailed recording in this electrode potential range permits the determination of the true electrode surface area and the roughness factor R_f. The area in the plotted CV under the recorded curve is determined in the range $0.0 < E_{RHE} < 0.36$ V. This area corresponds to the charge consumed for the hydrogen adsorption, and the double-layer charging; calculation of the charge (in A·s) depends on the setting of the used instruments and the scan rate. The charge Q_H^- needed for formation of H_{ad} in this potential range is obtained from the respective double-layer charge Q_{DL}, i.e. 2.25 times the charge obtained from the area under the curve in the CV in the range $0.48 < E_{RHE} < 0.64$ V.

From the CV displayed in Fig. 3.18 the charge in the range $0.0 < E_{RHE} < 0.36$ V is

$$Q_H^- = 1.68 \text{ mA} \cdot s = 1.68 \cdot 10^{-3} \text{ A} \cdot s \tag{3.69}$$

The double-layer charge Q_{DL} is

$$Q_{DL} = 0.1 \text{ mA} \cdot s = 0.1 \cdot 10^{-3} \text{ A} \cdot s \tag{3.70}$$

The charge actually needed for formation of H_{ad} is

$$Q_H^- = 1.68 \text{ mA} \cdot s - 0.1 \text{ mA} \cdot s = 1.58 \cdot 10^{-3} \text{ A} \cdot s \tag{3.71}$$

Assuming $1.3 \cdot 10^{15}$ platinum atoms per cm^2 on an ideally smooth surface and that in the range $0.0 < E_{RHE} < 0.36$ V 90% of all platinum atoms carry a hydrogen atom (this is equivalent to a degree of coverage of $\theta = 0.9$) with the elementary charge $q_e = 1.6 \cdot 10^{-19}$ A·s, a charge of $2.1 \cdot 10^{-4}$ A·s·cm^{-2} is needed for the coverage of an ideally smooth surface with adsorbed hydrogen. The charge Q_H^- implies a surface area of $1.58 \cdot 10^{-3}$ A·s$/2.1 \cdot 10^{-4}$ A·s·cm$^{-2} = 7.5$ cm^{-2}. The roughness factor is the ratio of true versus geometric electrode surface area. With a geometric surface area of the used platinum electrode of 3 cm^{-2} the roughness factor $R_f = 4$ can be calculated.

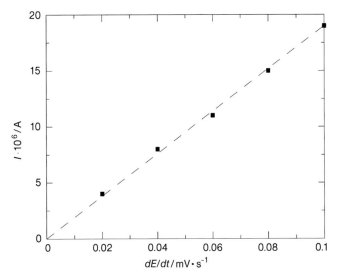

Fig. 3.21 Plot of current in the double-layer region as a function of scan rate for a platinum electrode of 3 cm^{-2} geometric surface area in an aqueous electrolyte solution of 0.05 **M** H$_2$SO$_4$, nitrogen purged.

Potential range $0.35 < E_{RHE} < 0.6$ V (double-layer region)

At $E_{RHE} = 0.6$ V the currents observed at the various scan rates are obtained. A plot of these currents vs. the scan rate yields Fig. 3.21; the behavior of the double layer closely resembles that of a plate condenser.

The slope in the interpolated line is equivalent to the double-layer capacity C_{DL}. The actual value is $C_{DL} = 194$ µF with a geometric surface area 3 cm^{-2}. With a double-layer capacity for an ideally smooth surface $C_{DL} = 20$ µF·cm^{-2}, a roughness factor $R_f = 194$ µF/20 µF·cm$^{-2} = 3.2$ is obtained, close to the value obtained from hydrogen adsorption measurements.

2. CV of formic acid

The plot shows current peaks in the positive-going scan at $E_{RHE} = 0.6$, 1.0, and 1.5 V. In the negative-going scan only a single wave close to the first wave of the positive going scan is observed. The peaks at $E_{RHE} = 1.0$ and 1.5 V are assigned to the formation of Pt-OH- and PtO-chemisorption layers. The wave at $E_{RHE} = 0.6$ V in the negative-going scan is explained as being due to the oxidation of formic acid after reduction of the Pt-O coverage (see the small reduction current peak around $E_{RHE} = 0.75$ V) on the bare, now again reactive platinum surface.

3. CV of a nickel electrode

Figure 3.22 shows two consecutive CVs of a nickel wire electrode in contact with an aqueous electrolyte solution of 0.05 **M** H$_2$SO$_4$. Before the first scan the electrode was kept at $E_{RHE} = -0.4$ V for ten seconds in order to reduce any conceivably present oxide layer residues or passivating layers.

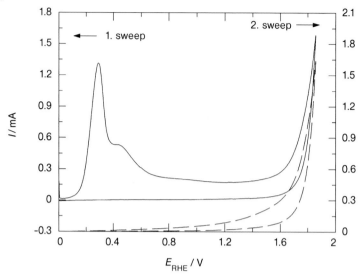

Fig. 3.22 CV of a nickel electrode in an aqueous electrolyte solution of 0.05 **M** H_2SO_4, nitrogen purged; $dE/dt = 0.1$ V·s^{-1}; second curve offset for clarity.

The passivation potential can be easily identified as the current maximum in the first scan. A Flade potential E_F, defined as the potential where the anodic current drops to a very low value in the passive region, cannot be identified easily with this system. The anodic current rising at potentials at the upper end of the passive potential region in the transpassive region is caused by further oxidation of the nickel surface and oxygen evolution. An activation potential (this potential is sometimes also called E_F), where the passivated electrode is returned to the active state (of dissolution) in the negative-going scan, cannot be observed in the system studied here. In case of electrodes of noble metals showing no passivation a breakthrough potential E_B can be obtained by extrapolation of the current rise in the positive-going scan towards the x-axis.

Questions
- Is cyclic voltammetry a stationary technique?
- Can redox potentials E_{red} (formal potentials E_0) of dissolved substances be derived from CVs?
- How do you distinguish between geometric area and true surface area?

Literature

Uhlig's corrosion handbook (R. W. Revie, H. H. Uhlig, Eds.), Wiley, New York 2000.
P. R. Roberge: Corrosion Basics: An Introduction (2nd ed.), NACE International, Houston, Texas, USA 2006.
H. Kaesche: Corrosion of metals, Springer, Berlin 2003.
H. Gerischer, Angew. Chem. **70** (1958) 285.

Experiment 3.12: Slow Scan Cyclic Voltammetry

Task

The electrochemical behavior of the redox systems Pb/Pb^{2+} and Pb^{2+}/Pb^{4+} is studied with slow scan rate CV.

Fundamentals

Cyclic voltammetry can be performed potentiostatically with scan rates set within a very wide range of possible values. According to the designation of this method as a non-stationary one (sometimes it is also called quasi-stationary) a very low scan rate approaches the stationary case, whereas a high scan rates results in an unstable response of the studied system. In this experiment a very low scan rate is employed; the results thus resemble those obtained during stepwise measurements of current density vs. electrode potential curves by stepping the electrode potential in small increments and recording the caused current after it has reached a stationary value.

For the study of the present redox systems two electrolyte systems are suggested resulting in extremely different responses. This illustrates again the need for the definition of an electrode first proposed by W. Nernst as a combination of an electron-conducting material in contact with an ionic solution (e.g., a lead wire and sulfuric or perchloric acid). In perchloric acid lead ions are very soluble (at $T=25\,°C$ 81 wt% of $Pb(ClO_4)_2$ in water), whereas in sulfuric acid the poorly soluble lead sulfate (at $T=25\,°C$ 0.0084 wt% of $PbSO_4$ in water) controls the behavior. In the first electrolyte solution an electrode of the first kind (sometimes also called solution electrode) is established, and in the sulfuric acid solution a electrode of the second kind is formed.

In perchloric acid (system I) the following redox reactions proceed:

$$Pb^{2+} + 2\ e^- \leftrightarrows Pb \tag{3.72}$$

$$Pb^{4+} + 2\ H_2O \leftrightarrows PbO_2 + 4\ H^+ + 2\ e^- \tag{3.73}$$

The use of a platinum sheet instead of a lead foil enables the second redox reaction to be studied; lead itself would not be passivated in perchloric acid but would be completely dissolved anodically.

In aqueous sulfuric acid (system II) the processes well known from the lead acid battery (starter battery in vehicles, see Expt. 6.1 and EC:441) occur:

$$PbSO_4 + 2\ e^- \leftrightarrows Pb + SO_4^{2-} \tag{3.74}$$

$$PbSO_4 + 2\ H_2O \leftrightarrows PbO_2 + 4\ H^+ + 2\ SO_4^{2-} \tag{3.75}$$

For this experiment commercial lead [12] sheet or foil can be used as an electrode.

---

12) Sometimes commercial lead may contain alloying components which are electrochemically active themselves, and this might cause unexpected experimental artifacts. When no sufficiently pure lead is available a piece of an electrode from a lead acid battery may be used; alloying metals employed therein will not cause artifacts.

Execution

Chemicals and instruments

Aqueous solution of sulfuric acid 1 M

Aqueous solution of $Pb(ClO_4)_2$ 1 M

Mixture (1:1) of glacial acetic acid and aqueous solution of hydrogen peroxide (30%) (to remove traces of PbO_2 from platinum sheets)

Potentiostat

Triangular voltage sweep generator

X-Y-recorder (as an alternative: PC with interface card)

2 Platinum electrodes

2 Lead electrodes

PTFE tape

Mercurous sulphate and lead wire reference electrode

Three-electrode cell (H-cell)

Setup

With respect to the numerous conceivable combinations of instruments (potentiostats, function generators, recorders, PCs) a scheme of the wiring of instruments does not seem to be necessary.

For experiments with system I (perchloric acid) platinum sheets are used as working and counter electrode; a lead wire serves as reference electrode. At this wire in contact with the electrolyte solution containing $Pb(ClO_4)_2$ the redox potential of the Pb/Pb^{2+}-electrode is established.

With system II a piece of lead sheet is used as working electrode. PTFE tape is wrapped around the metal leaving a well-defined surface area as active electrode surface. As counter electrode a piece of lead prepared in the same way or the lead wire from the previous experiment is used. Instead of the suggested mercurous sulphate reference electrode a hydrogen reference electrode can be used; electrode potential values given below have to be converted in this case.

Procedure

In the electrode potential range $- 10 < E < 2000$ mV a CV is recorded at $dE/dt = 1$ mV·s^{-1}. Both expected redox processes are expected to proceed in this potential window. If necessary the potential limits have to be adjusted somewhat. Observed current densities should not exceed $j = 100$ mA·cm^{-2}. If a measurement of the transferred electrical charges (e.g., by measuring the area in a CV) is desired, separate CVs of only one redox process should be recorded in a narrower electrode potential range.

With system II and a mercurous sulphate reference electrode the electrode potential range should be $-1600 < E < 1600$ mV.

Evaluation

Figure 3.23 shows a complete CV with system I. Integration of the areas under the CV trace associated with anodic lead dissolution and cathodic lead deposition yields the respective charges. Care must be exercised when determining the charge of lead deposition: The area under both the negative- and the positive-going scan must be measured. As a result the lead electrode (the redox system Pb/Pb^{2+}) turns out be very reversible; anodic and cathodic charge are equal. The current potential curves in the CV pass the axis with a large slope indicative of fast electrode kinetics and a large exchange current density ($j_{00} \approx 100 \, \text{A} \cdot \text{cm}^{-2}$). The corresponding evaluation of the Pb^{2+}/Pb^{4+} redox system suggests that this process is much less ideal, i.e. formation of PbO$_2$ requires a considerable overpotential. The much smaller slope implies a much smaller exchange current density ($j_{00} \approx 1 \, \text{mA} \cdot \text{cm}^{-2}$).

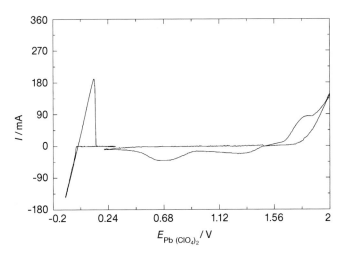

Fig. 3.23 CV of a platinum electrode in an aqueous electrolyte solution of 1 **M** Pb(ClO$_4$)$_2$, dE/dt=2 mV·s^{-1}.

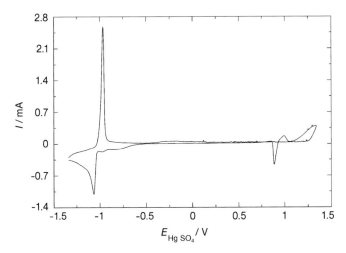

Fig. 3.24 CV of a lead electrode in an aqueous electrolyte solution of 1 **M** H$_2$SO$_4$, dE/dt=2 mV·s^{-1}.

Figure 3.24 shows a complete CV with system II. Lead dissolution and deposition proceed with high yield. At the redox couple Pb^{2+}/Pb^{4+} formation of lead dioxide and of oxygen can hardly be separated. A comparison of the charge consumed in the reduction of PbO_2 and that consumed in the formation of $PbSO_4$ (at the Pb/Pb^{2+} redox couple) suggests nevertheless the essential reversibility of the Pb^{2+}/Pb^{4+} redox couple. The anodic current peak observed in the negative-going potential scan has been assigned to the oxidation of water (yielding oxygen) by Pb^{3+} ions formed as intermediates.

Literature
F. Beck in: The Electrochemistry of Lead (A. T. Kuhn ed.), Academic Press, New York 1979, p. 65.
J. G. Sunderland, J. Electroanal. Chem. **71** (1976) 341.

Experiment 3.13: Kinetic Investigations with Cyclic Voltammetry

Task
The exchange current density j_0 and the symmetry coefficient a of a redox system are determined by cyclic voltammetry.

Fundamentals
Cyclic voltammetry can be employed to determine kinetic data of an electrode charge transfer reaction (exchange current density j_0 and the symmetry coefficient a) provided that mass transport to a stationary (i.e. not moving) electrode in a solution (also not moving, e.g., stirred) can be treated mathematically. Mass transfer will proceed solely by diffusion (i.e. driven by concentration gradients) in this case. Details of the rather complicated mathematics can be found in the literature [13]. In this experiment only the relationships between observables and these kinetic data will be considered.

A cyclic voltammogram (CV) obtained with a metal electrode in contact with an electrolyte solution containing a redox system is shown below (Fig. 3.25).

Characteristic parameters for a simple CV of a redox system

$$Ox^+ + e^- \rightarrow Red^0 \tag{3.76}$$

are
1. Height of the current peaks I_p
2. Potential difference between the peak potentials ΔE_p.

The mathematical treatment provides the relationships between I_p (or the peak current density j_p), the peak potential difference ΔE_p, the scan rate $v = dE/dt$, the kinetic parameters exchange current density j_{00}, and the symmetry coefficient a.

---

13) See references at the end of the description of this experiment.

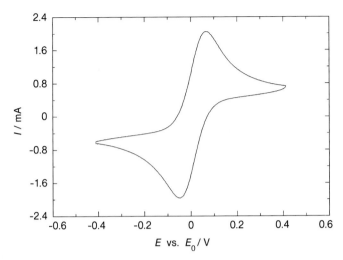

Fig. 3.25 CV of the redox system Fe^{2+}/Fe^{3+}, single cyclic scan at $dE/dt = 50$ mV·s^{-1}, platinum electrode in an aqueous electrolyte solution of 5 **mM** each of $Fe(NH_4)(SO_4)_2\cdot12\ H_2O$ and $FeSO_4\cdot7\ H_2O$ and 0.5 **M** H_2SO_4.

In the case of a relatively slow[14] electrode reaction the peak current of the cathodic peak depends on the scan rate according to:

$$I_p = 3.01 \cdot 10^5 \cdot A \cdot n^{3/2} \cdot (1 \cdot a)^{1/2} \cdot D_{ox}^{1/2} \cdot c_{0,ox} \cdot \sqrt{v} \tag{3.77}$$

A plot of I_p vs. $v^{1/2}$ yields a line, and the slope enables calculation of a. Calculation for an oxidation step proceeds in the same manner. The rate constant k_0 can be derived from the peak potential difference ΔE_p, and finally the exchange current density can be the calculated

$$j_0 = F \cdot k_0 \cdot c_{ox}^a \cdot c_{red}^{(1-a)} \tag{3.78}$$

From ΔE_p a value Y can be derived (see Table 3.1).

Because the relationship cannot be given numerically Y may also be obtained with a graph (Fig. 3.26). The rate constant can be obtained from Y:

---

14) This case is frequently called an irreversible reaction, whereas a fast reaction is called reversible. Unfortunately use of the terms reversible and irreversible in chemistry (including electrochemistry) is rather confused. In thermodynamics the use is well defined, and when used in studies of reaction mechanisms the term reversible describes a reaction proceeding in both reactions (forward and backward) along the same path only in opposite directions. This use apples also to electrode reactions (e.g., the redox system studied here). In addition the term reversible is applied to the rate of charge transfer reactions when the rate is high enough to keep the ratio of reactants (redox component) at the electrode surface at the values given by the Nernst equation. In case of an irreversible reaction establishment of this ratio lags behind.

$$Y = (D_{ox}/D_{red})^{a/2} \cdot k_0((R \cdot T)^{1/2}/(n \cdot F \cdot v \cdot D_{ox})^{1/2}) \tag{3.79}$$

Table 3.1

Y	ΔE_p/mV	Y	ΔE_p/mV
20	61	1	84
7	63	0.75	92
5	65	0.5	105
4	66	0.35	121
3	68	0.25	141
2	72	0.1	212

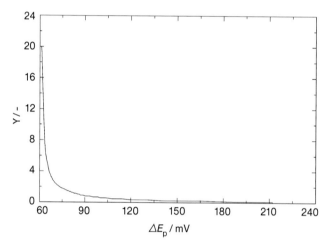

Fig. 3.26 Plot of Y vs ΔE_p.

The potential limits in a CV experiment are mostly either the potentials of onset of oxygen or hydrogen evolution respectively or potentials sufficiently positive and negative with respect to the redox peak pair.

Execution
Chemicals and instruments
Aqueous solution of sulfuric acid 0.5 **M**
Aqueous solution of Fe(NH$_4$)(SO$_4$)$_2 \cdot 12$ H$_2$O 0.05 **M** (stock solution)
Aqueous solution of FeSO$_4 \cdot 7$ H$_2$O 0.05 **M** (stock solution)
Nitrogen purge gas
Potentiostat
Triangular voltage sweep generator
X-Y-recorder (as an alternative: PC with interface card)
3 Platinum electrodes
Three-electrode cell (H-cell)

Setup

With respect to the numerous conceivable combinations of instruments (potentiostats, function generators, recorders, PCs) a scheme of the wiring of instruments does not seem to be necessary. Instead of a reference electrode simply a platinum electrode is used. At this electrode immersed in the electrolyte solution containing both components of the investigated redox system, the rest potential E_0 of the redox system will be established; this potential is used as a point of reference throughout this experiment.

Procedure

In this experiment the redox system Fe^{2+}/Fe^{3+} at a platinum electrode is studied in an aqueous solution 0.05 **M** of sulfuric acid. The respective diffusion coefficients of the ions are

$$D_{red} = 5.04 \cdot 10^{-6} \text{ cm}^2 \cdot \text{s}^{-1}$$
$$D_{ox} = 4.65 \cdot 10^{-6} \text{ cm}^2 \cdot \text{s}^{-1}$$

The cell is filled with an aqueous solution of 5 **mM** each of $Fe(NH_4)(SO_4)_2 \cdot 12$ H_2O, $FeSO_4 \cdot 7$ H_2O, and 0.5 **M** H_2SO_4 prepared by mixing adequate amounts of sulfuric acid and the stock solutions. Purging with argon or nitrogen for about 10 min removes traces of oxygen dissolved in the electrolyte solution. CVs are recorded at different scan rates ($dE/dt=0.1$, 0.2, 0.4, 0.6, 0.8, and 1.0 $mV \cdot s^{-1}$). Typical results are shown below.

Evaluation

A plot of peak currents vs. scan rate yields a line as shown in Fig. 3.28.

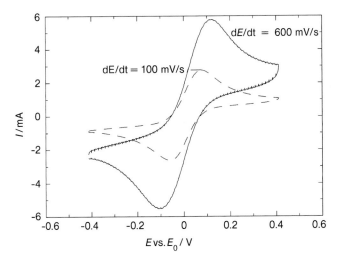

Fig. 3.27 CVs of the redox system Fe^{2+}/Fe^{3+}, single cyclic scan at different scan rates, platinum electrode in an aqueous electrolyte solution of 5 **mM** each of $Fe(NH_4)(SO_4)_2 \cdot 12$ H_2O and $FeSO_4 \cdot 7$ H_2O and 0.5 **M** H_2SO_4, nitrogen purged.

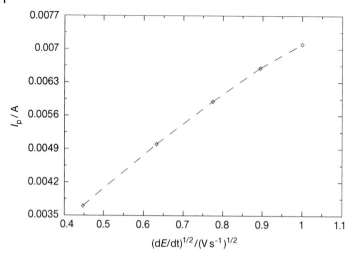

Fig. 3.28 Plot of peak currents vs. $(dE/dt)^{1/2}$ for the redox system of Fig. 3.27.

From the slope a of the line the symmetry coefficient α can be calculated according to

$$1 - \alpha = \left\{ \frac{a}{3.01 \cdot 10^5 \cdot A \cdot n^{3/2}(4.65 \cdot 10^{-6})^{1/2} \cdot c_{0,\text{ox}}} \right\}^2 \tag{3.80}$$

If peak current densities instead of peak currents (as here) have been obtained, the electrode area is dropped from the equation. The mathematical relationship results in a high sensitivity of the results vs. exact concentration values[15] or other errors. In the example shown here $\alpha = 0.55$. In the case of extremely rough electrodes the roughness factor as determined in the preceding experiment should be used to take into account the true surface area of the electrode.

From the data of Table 3.1 a calibration curve (as shown in Fig. 3.26) is drawn. From the CVs ΔE_p corresponding values of Y are obtained: they are used to calculate k_0 and j_{00}. From the CVs shown in part in Fig. 3.27 the following results are obtained:

---

15) Concentrations per cm^3 have to be entered, in this example
$c_{0,\text{ox}} = 5 \cdot 10^{-6} \ \text{Mol} \cdot \text{cm}^{-3}$.

Table 3.2 Results of cyclic voltammetry

$dE/dt/V \cdot s^{-1}$	$\Delta E_p/V$	Y	$k_0/cm \cdot s^{-1}$
0.2	0.115	0.45	0.002840
0.4	0.140	0.224	0.001999
0.6	0.145	0.224	0.002448
0.8	0.150	0.224	0.002827
1.0	0.180	0.14	0.001976

The average value of $k_0 = 0.002418$ cm·s^{-1}, the exchange current density is $j_0 = 1.1$ mA·cm^{-2}.

Literature

R. S. Nicholson, Anal. Chem. **37** (1965) 1351.

A. J. Bard and L. R. Faulkner: Electrochemical Methods, Wiley, New York 2001, p. 226.

Experiment 3.14: Numerical Simulation of Cyclic Voltammograms

Task

The kinetic data j_0 and a of a redox system are determined by numerical simulation of experimental cyclic voltammograms.

Fundamentals

Kinetic data (exchange current density j_0 and symmetry coefficient a) of a redox system can be determined by numerical simulation of experimentally obtained cyclic voltammograms. The extensive mathematical fundamentals of numerical simulation are described in the literature. At the time of writing this book numerous programs for simulation and for more or less automatic fitting of simulations to experimental data were available as free public domain software, shareware and commercially available software. Because availability, prices, and modes of licensing are in permanent change no attempt is made to provide a list of programs. Results shown here were obtained with the program Polar 4.3 (Dr. Huang Pty Ltd). Simulation and fitting are also possible with software supplied with an introductory book: D. K. Gosser Jr.: Cyclic Voltammetry, VCH, New York 1993.

Execution

Chemicals and instruments

Aqueous solution of 5 **mM** $K_3Fe(CN)_6$ + 0.5 **M** K_2SO_4

Potentiostat

Triangular voltage sweep generator

X-Y-recorder (as an alternative: PC with interface card)

2 Platinum electrodes
Saturated calomel reference electrode
Three-electrode cell (H-cell)

Setup

With respect to the numerous conceivable combinations of instruments (potentiostats, function generators, recorders, PCs) a scheme of the wiring of instruments does not seem to be necessary. Instead of the saturated calomel reference
electrode simply a platinum electrode can be used. At this electrode immersed
in the electrolyte solution containing both components of the investigated redox
system the rest potential E_0 of the redox system will be established, this potential is used as a point of reference throughout this experiment. In this case the
formal potential E_0 of the redox potential is not determined within the evaluation as described below.

Procedure

By numerical simulation of the measured CVs, kinetic data are obtained. Typical
examples are shown in Figs. 3.29 and 3.30. Into the simulation the carefully determined electrode surface areas, the supporting electrolyte and redox system
concentration, and the sweep rate are entered. The value of E_0 is initially estimated; the exact value of $E_{0,SCE}=0.235$ V is determined within the simulation
procedure.

The kinetic data are $k_0=3.1\cdot10^{-3}$ cm$\cdot$s^{-2} and $a=0.5$.

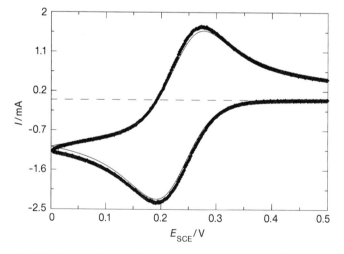

Fig. 3.29 Simulated ($-$) and measured CV at $dE/dt=0.1$ V$\cdot$s^{-1}.

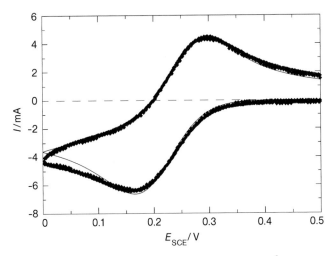

Fig. 3.30 Simulated (—) and measured CV at $dE/dt = 1 \ V \cdot s^{-1}$.

Experiment 3.15: Cyclic Voltammetry with Microelectrodes

Task

With a microelectrode CVs are recorded at various scan rates; these are compared with expectations based on various models of transport.

Fundamentals

At electrodes of common shape and size (e.g., disc electrodes of some millimeters in diameter, metal sheet electrodes of a few square centimeters in size) used as working electrodes in electrochemistry, planar diffusion of reactants can be safely assumed. The extremely small size of a mercury drop in polarography, in particular its extremely rounded spherical surface, cause spherical diffusion. A similar situation is found at the phase boundary electrode|solution, when the active surface area of the electrode has a characteristic dimension (disc diameter, width of a strip) of the same order of magnitude as the diffusion layer thickness. At typical experimental conditions these are a few micrometers. At such microelectrodes spherical diffusion proceeds also. Microelectrodes can be prepared by embedding thin metal or carbon wires or fibers (of a few micrometers diameter) into inert materials (glass, resin). The active electrode surface may be embedded with its surface level with the surrounding surface, but also protruding hemispheres or small spheres attached to thin wires etc. are conceivable.

At microelectrodes the laws of planar or linear diffusion are valid no longer. Fig. 3.31 shows mass transport according to linear diffusion to a "large" electrode and spherical diffusion to a microelectrode (EC:201). This change in mode of diffusion also causes a considerable change in the shape of a CV because the electrode size is of the typical dimensions of the thickness of the Nernst diffusion layer.

Fig. 3.31 Mass transport at a planar ("large") electrode (left) and at a microelectrode (right).

This figure implies that mass transport to a microelectrode is composed of a flux according to planar diffusion I_p and an additional contribution according to spherical diffusion I_{sph}. In the diffusion-limited case both fluxes are added together

$$I = I_p + I_{sph} \tag{3.81}$$

With the radius r of a microelectrode (a circular disc embedded in an insulating material is assumed) and a form factor a the current is

$$I_{sph} = a \cdot r \cdot n \cdot F \cdot D \cdot c \tag{3.82}$$

The form factor for a flat disc is $a=4$; for a sphere $a=4\pi$ and for a hemisphere $a=2\pi$. The relative contributions from both currents depend on the ratio of the typical electrode dimension r_0 to the diffusion layer thickness. As a criterion with time t the quotient $D \cdot t / r_0^2$ can be used. At a value large than 1, i.e. with a diffusion layer thickness substantially larger than r_0, the current reaches a constant limiting value which can be observed easily in a CV. In the inverse case the typical shape of a CV with anodic and cathodic current peaks can be ob-

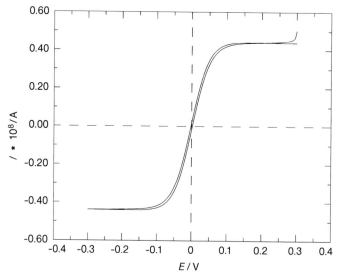

Fig. 3.32 Simulated CV of a microelectrode with $r_0 = 0.001$ cm at $dE/dt = 0.01$ V·s^{-1}.

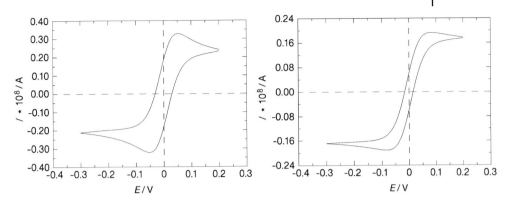

Fig. 3.33 Simulated CVs of a microelectrode with: (left) $r_0=0.01$ cm at $dE/dt=0.01$ V·s^{-1}, (right) $r_0=0.01$ cm at $dE/dt=0.001$ V·s^{-1}.

served. Because of the constant electrode potential scan rate v the electrode potential E is equivalent to the parameter t in the quotient used as the criterion. Thus both cases can be demonstrated easily with an electrode of fixed dimension r_0 at various scan rates. Fig. 3.32 shows simulated CVs with a disc electrode of diameter $r_0=0.001$ cm. The influence of scan rate and electrode diameter can easily e illustrated. Increase of the electrode diameter by an order of magnitude results at the same scan rate as before in a CV of traditional shape (see Fig. 3.33), with this diameter and a scan rate lower by an order of magnitude, again limiting currents can be observed. These effects are examined qualitatively in this experiment.

Execution
Chemicals and instruments
Aqueous solution of **5 mM** $K_3Fe(CN)_6$ + **5 mM** $K_4Fe(CN)_6$ + **0.5 M** K_2SO_4
Nitrogen purge gas
Potentiostat
Triangular voltage sweep generator
X-Y-recorder (as an alternative: PC with interface card)
Microelectrode (for preparation see p. 5)
2 Platinum electrodes (counter and reference electrode)
Three-electrode cell (H-cell)

Setup
The standard setup for cyclic voltammetry (see preceding experiments) is used.

Procedure
With the electrolyte solution saturated with nitrogen gas CVs are recorded at different scan rates.

Evaluation

Figs. 3.34 and 3.35 show CVs obtained with a carbon fiber electrode embedded in epoxy resin inside a glass capillary at various scan rates. At low scan rates the typical CV of a microelectrode with diffusion-limited currents is observed; at a higher scan rate in the same setup a conventional CV is recorded.

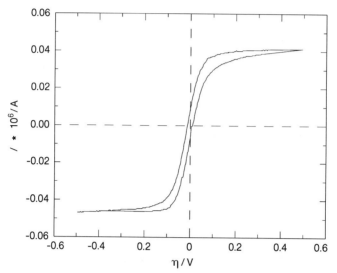

Fig. 3.34 CV of a microelectrode in an aqueous solution of 5 **mM** $K_3Fe(CN)_6 + 5$ **mM** $K_4Fe(CN)_6 + 0.5$ **M** K_2SO_4 at $dE/dt = 0.005$ V·s^{-1}.

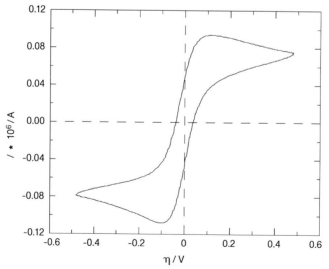

Fig. 3.35 CV of a microelectrode in an aqueous solution 5 **mM** $K_3Fe(CN)_6 + 5$ **mM** $K_4Fe(CN)_6 + 0.5$ **M** K_2SO_4 at $dE/dt = 0.1$ V·s^{-1}.

Experiment 3.16: Cyclic Voltammetry of Organic Molecules

Task
With cyclic voltammetry the reactions of *N,N*-dimethylaniline and 2,6-dimethyl-aniline are investigated, and reaction pathways for both electrochemical and chemical reactions are derived.

Fundamentals
Beyond the already described applications of cyclic voltammetry in studies of electrode kinetics, determination of formal potential E_0, determination of double-layer capacities, and true electrode surface areas and electrode potentials where conversion of dissolved or adsorbed species occur by charge transfer, this method has also been established as a powerful tool in mechanistic studies of complicated reaction sequences involving charge-transfer step(s) coupled with preceding, follow-up, or otherwise coupled homogeneous or heterogeneous chemical reactions. This intriguing potential has helped to establish CV as a common method in organic as well as inorganic, and in particular organometallic and coordination, chemistry. Electron transfer to coordinated metal ions as well to their ligands provides insight into modes of bonding, reactivities and intra- as well as intermolecular interaction.

Depending on the type of investigation a single potential scan starting at an initial potential defined either by the task or by the system's properties to a final potential or a single or multiple cyclic scan from the starting potential to the potential limit and back to the starting point [16] may be performed. Whereas in

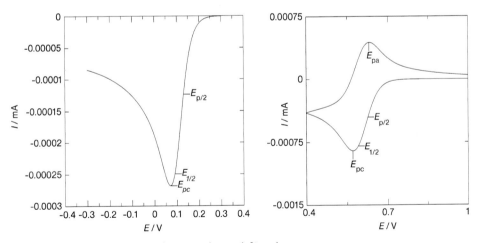

Fig. 3.36 Simulated CVs for a single potential scan (left) and a single cyclic scan with typical values indicated.

---

16) In cyclic experiments these potentials are more suitably called the upper (anodic) and lower (cathodic) potential limit.

the preceding experiments only peak potentials and peak currents were of importance in the studies described, the following further details are meaningful These are indicated in the simulated CVs (Fig. 3.36).

In a discussion peak potential E_{pa} and E_{pc} and the corresponding peak currents I_a and I_c are of primary interest. The so-called half-wave potentials $E_{1/2}$ (so called in analogy with polarography, see Expt. 4.8) are in most cases half-peak potentials $E_{p/2}$ [17]. The mathematical relation is relatively simple; the application requires knowledge of the electrode reaction:

$$E_{p/2} = E_{1/2} + 1.09 \frac{R \cdot T}{n \cdot F} = E_{1/2} + 0.028/n \; [\mathrm{mV}] \tag{3.83}$$

If a species formed by electron transfer in the first reduction is subject to a further reduction (EE-mechanism [18]) the CV depicted in Fig. 3.37 is observed.

When a chemical reaction follows the first electron transfer (ECE mechanism) in the CV shows a substantial change as seen in Fig. 3.38.

Because species formed in the first reduction step are converted in a following chemical reaction into species reduced in the second current wave in the anodically-going scan, these reduced species can be reoxidized again, but no second anodic wave or a very weak one only appears. This is caused by an almost complete lack of species of the first reduction step – these species have been consumed in the chemical reaction.

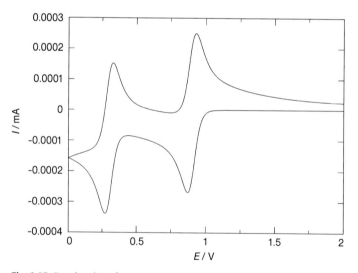

Fig. 3.37 Simulated CV for a cyclic potential scan with two consecutive electron transfers.

17) or worse: they are really E_0.
18) Reaction sequences are described symbolically with letters E referring to an electron transfer and C referring to a chemical reaction.

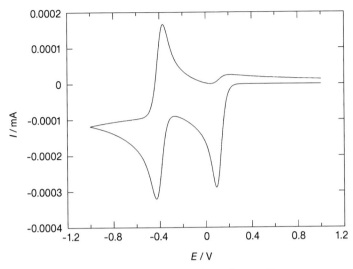

Fig. 3.38 Simulated CV for a cyclic potential scan for an ECE mechanism.

Analysis of the second peak as a function of scan rate may provide information about the rate of the chemical reaction. With increasing time this peak will increase, because less time remains for consumption of the products of the first reduction by the chemical reaction. In turn this might also cause a change of the peak ratio of the first and the second reduction peak, because the second reduction depends on the chemical reaction following the first reduction.

Execution
Chemicals and instruments
Aqueous solution of 2 **mM** *N,N*-dimethylaniline + 0.5 **M** H_2SO_4
Aqueous solution of 10 **mM** 2,6-dimethylaniline + 0.5 **M** H_2SO_4
Nitrogen purge gas
Potentiostat
Triangular voltage sweep generator
X-Y-recorder (as an alternative: PC with interface card)
Hydrogen reference electrode
2 Platinum electrodes
2 Gold electrodes
Three-electrode cell (H-cell)

Setup
The standard setup for cyclic voltammetry (see preceding experiments) is used.

Procedure

1. Oxidation of N,N-dimethylaniline

Starting at $E_{RHE}=0.4$ V a series of CVs with an upper potential limit $E_{RHE}=1.14$ V at $dE/dt=0.1$ V·s^{-1} and a platinum electrode is recorded.

2. Oxidation of 2,6-dimethylaniline

Starting at $E_{RHE}=0.4$ V a series of CVs with an upper potential limit $E_{RHE}=1.14$ V at $dE/dt=0.1$ V·s^{-1} and a gold electrode is recorded.

Evaluation

1. Oxidation of N,N-dimethylaniline (Fig. 3.39)

In the first anodic-going scan an anodic peak caused by oxidation of N,N-di-methylaniline at approx. $E_{RHE}=1.1$ V is observed.

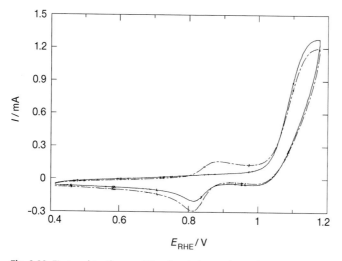

Fig. 3.39 First and tenth scan CVs of a platinum electrode in an aqueous solution of 2 **mM** N,N-dimethylaniline + 0.5 **M** H$_2$SO$_4$ at $dE/dt=0.1$ V·s^{-1}.

In the cathodic-going scan a peak is found around $E_{RHE}=0.82$ V. The tempting assignment of this peak to a reduction of the previously formed oxidation products is untenable, because an anodic wave corresponding to this reduction wave is formed around $E_{RHE}=0.89$ V. Both latter peaks increase in height during further cycling. Considerations of conceivable reaction products of the initially formed radical cation suggest formation of N,N,N',N'-tetramethyl benzidine. The light yellow coloration of the electrolyte solution supports this proposal. It can be proved by recording a CV of this compound that the peak pair found here is indeed observed again. The respective reaction equation is shown in Fig. 3.40.

This proposal can be supported with a simulated CV (Fig. 3.41).

For this simulation a species present in its reduced form is assumed. It is electro-oxidized at $E=0$ V. This product is converted chemically into a compound (ECE mechanism) electro-active with a redox potential $E_0 = -0.4$ V. In the first scan only the reduction wave is observed, and in the second scan the reoxidation can also be observed. According to the consumption of the starting material the large anodic peak is substantially diminished.

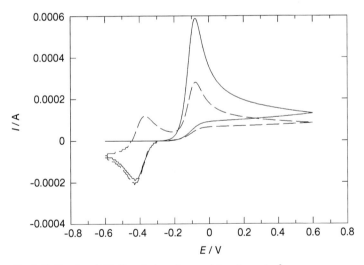

Fig. 3.40 Reaction sequence of the conversion of *N,N*-dimethylaniline and its follow-up products.

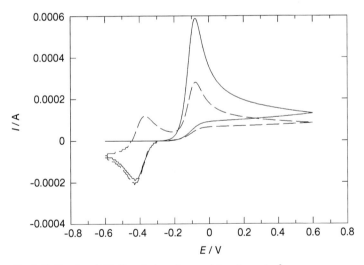

Fig. 3.41 Simulated CV (first (—) and second (– – –) scan) of an ECE mechanism; for further details see text.

2. Oxidation of 2,6-Dimethylaniline (Fig. 3.42)
The behavior of this compound is significantly different.

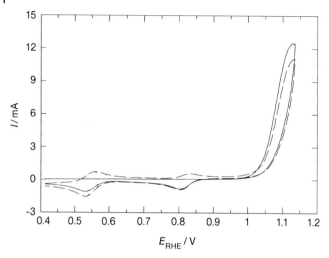

Fig. 3.42 First and second scan CVs of a gold electrode in an aqueous solution of 10 **mM** 2,6-dimethylaniline + 0.5 **M** H_2SO_4 at $dE/dt = 0.1$ V·s^{-1}.

In the first scan anodic oxidation of the starting material is observed, and this oxidation product is reduced in two subsequent reduction steps. In the second scan two oxidation peaks related to these reduction peaks are found, the large oxidation peak being somewhat diminished. Because both redox peak pairs show exactly the same behavior, after some stirring of the electrolyte solution they cannot be associated with oligo- or polymers attached to the electrode. The identity of the species causing these redox peaks can be verified by recording a CV of the suggested compounds (Most likely they are a substituted benzidine [19] formed by tail-to-tail coupling and p-phenylenediamine formed by head-to-tail coupling.)

Literature

B. Speiser, Curr. Org. Chem. **3** (1999) 171.
B. Speiser in: Electroanalytical Chemistry **19** (A. J. Bard, ed.), Marcel Dekker, New York 1996, p. 1.
T. Mizoguchi and R. N. Adams, J. Am. Chem. Soc. **84** (1962) 2058.
R. L. Hand and R. F. Nelson, J. Am. Chem. Soc. **96** (1974) 850.

19) Unsubstituted benzidine is highly carcinogenic, the substituted benzidines are less dangerous. Nevertheless they should be handled with care.

Experiment 3.17: Cyclic Voltammetry in Nonaqueous Solutions

Task
The redox electrochemistry of ferrocene is studied in a nonaqueous electrolyte solution with cyclic voltammetry.

Fundamentals
Cyclic voltammetry is used with increasing frequency to study processes in aqueous and also in nonaqueous solutions. Although the electrochemical fundamentals are the same for both types of solvents and solutions some instrumental and experimental details merit particular attention. The choice of suitable reference electrodes, careful exclusion of oxygen and moisture from the electrochemical cell, and thorough cleaning and drying of all substances and equipment are only the most obvious aspects. In this experiment the redox behavior of ferrocene at a platinum electrode in acetonitrile with tetraethylammoniumperchlorate as supporting electrolyte is studied. As well as the kinetics of the system, establishment of a reference potential is considered.

Execution
Chemicals and instruments
Acetonitrile (purified and dried[20])
Tetraethylammoniumperchlorate (dried)
Ferrocene
Purge gas (nitrogen)
2 Platinum electrodes
Silver chloride reference electrode filled with supporting electrolyte solution
Three-electrode cell (H-cell)
Setup for cyclic voltammetry

Setup
The standard setup for cyclic voltammetry (see preceding experiments) is used. The purge gas used to remove oxygen from the electrolyte solution should be piped into a fume hood because of the health risks of acetonitrile.

Procedure
The supporting electrolyte solution is prepared from the solvent and the electrolyte salt. This solution is also used to fill the reference electrode (already containing some pieces of solid silver chloride and the silver wire). In order to avoid contamination with ambient moisture swift working is recommended. The exact calculated amount necessary to obtain the desired concentration of ferrocene ($c_{\text{ferrocene}} = 1$ **mM**) is added to the main compartment of the cell. CVs are recorded at various scan rates.

---

20) For purification see p. 3.

Evaluation

Fig. 3.43 shows a typical set of CVs. A response as expected for a simple redox system is observed. The rather poor conductance of the electrolyte solution (substantially lower than of aqueous solutions) could pose a serious challenge to a potentiostat. The output voltage at the counter electrode required to establish the desired working electrode potential (compliance of the potentiostat) may be insufficient, resulting in poorly defined CVs especially at high scan rates. In addition, the signal-to-noise ratio is poorer, as can be seen in the somewhat noisy curves. The peak potential difference ΔE_p grows with scan rate indicating a slow electron transfer and a low exchange current. The potential value in the middle between respective peak potentials (also called th redox potential, formal potential or E_0) stays constant at $E_{Ag/AgCl} = 0.45$ V. Because the ferrocene/ferrocenium system is frequently used as a point of reference, its knowledge is helpful. With this value there is no need to run an additional CV with added ferrocene at the end of the study, employing cyclic voltammetry to establish a reference for the preceding measurements performed with the silver/silver chloride reference electrode previously employed. Any effects of the previously studied compounds still present in the electrolyte solution on the behavior of ferrocene are thus also excluded. Unfortunately the value given above is obviously valid only for the employed electrolyte solution because of solvent effects.

Literature

D. K. Gosser Jr.: Cyclic Voltammetry, VCH, New York 1993.

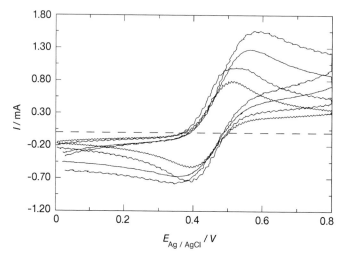

Fig. 3.43 CVs of a platinum electrode in a solution of 1 **mM** ferrocene in acetonitrile + 0.1 **M** Et$_4$NClO$_4$, dE/dt=0.05; 0.1; 0.5; 1 V·s^{-1}.

Experiment 3.18: Cyclic Voltammetry with Sequential Electrode Pocesses

Task

With cyclic voltammetry a reaction pathway for the electrooxidation of a highly substituted aromatic amine is derived.

Fundamentals

Radical cations A^{+*} are frequently formed as the product of electrooxidation of an organic molecule A. This reactive intermediate may react with various partners (solvent molecules, further molecules A, etc.); it can also be oxidized further, resulting in a dication A^{2+}. This in turn may be involved in further reactions. Because in all these cases chemical and reaction pathways compete, typical CVs can be obtained wherein the various oxidation potentials and rates of reactions will result in complicated interdependences between scan rate, potential limits and recorded CVs.

In the present case, considering only electrochemical reactions, the sequence

$$A \leftrightarrows A^{+*} + e^- \leftrightarrows A^{2+} + e^- \leftrightarrows B \tag{3.84}$$

will result in a simulated CV as shown below (Fig. 3.44).

With an anodic potential below the value needed to convert the radical cation into a dication the usual CV of a redox system is obtained. With a more positive potential limit, the latter reaction proceeds. Because of the fast chemical reaction of the dication assumed here, the reduction of the dication cannot be observed. With a slower chemical reaction or at a higher scan rate the reduction peak will be observed again. In both cases the reduction potential can be reached before the chemical reaction has consumed the chemical reaction product.

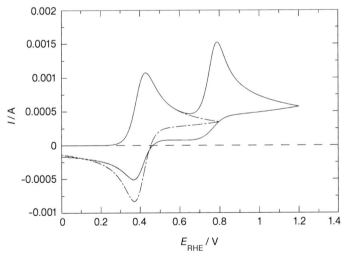

Fig. 3.44 Simulated CV for an EEC mechanism with different anodic potential limits; simulation with CVSIM (see Expt. 3.14).

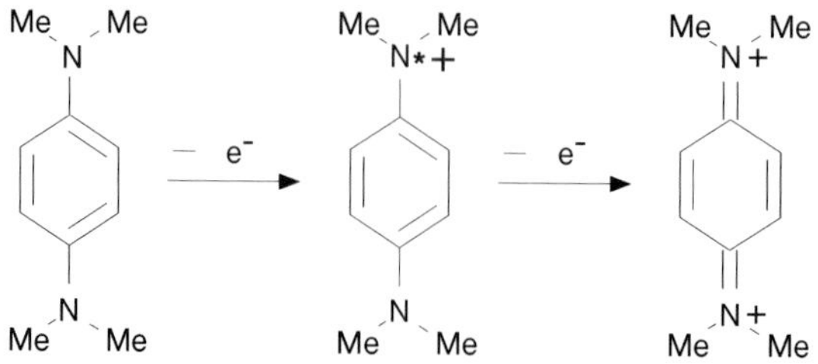

Fig. 3.45 Simplified reaction scheme of transformations of
N,N,N',N'-tetramethyl-p-phenylene diamine.

These possibilities can be explored taking N,N,N',N'-tetramethyl-p-phenylene
diamine as an example (Fig. 3.45). In aqueous solution the reversible electrooxidation yielding the radical cation[21] can be observed easily. Upon further oxidation the dication is formed which is consumed in aqueous solution by a chemical reaction (nucleophilic attack of solvent constituents).

Execution
Chemicals and instruments
Aqueous solution of 2 **mM** N,N,N',N'-tetramethyl-p-phenylene diamine in 0.5 **M**
H_2SO_4
Purge gas (nitrogen)
2 Platinum electrodes
Hydrogen reference electrode
Three-electrode cell (H-cell)
Setup for cyclic voltammetry

Setup
The standard setup for cyclic voltammetry (see preceding experiments) is used.

Procedure
CVs are recorded at different scan rates and with various anodic potential limits.

Evaluation
Figures 3.46 and 3.47 show typical CVs with various upper potential limits and
otherwise unchanged experimental conditions. For comparison CVs with the
supporting electrolyte solution only are also shown.
In Fig. 3.46 the expected redox peak pair is observed. The peak height ratio
implies (because of the somewhat lower height of the reduction peak) a compet-

---

21) This compound is called Wurster's blue
because of the intense blue color of the
oxidation product; it is used as Wurster's
reagent. The intense blue color of the oxidation product as well as the intense red

color of the oxidation product of N,N-dimethyl-
p-phenylene diamine were first observed by Ca-
simir Wurster (1856–1913), who assumed the
colored species to be iminium salts. In fact
they are radical cations.

ing chemical reaction consuming the radical cation. In Fig. 3.47 the formation of the dication is easily observed; the almost complete lack of a corresponding reduction wave indicates rapid consumption of the dication by a chemical reaction.

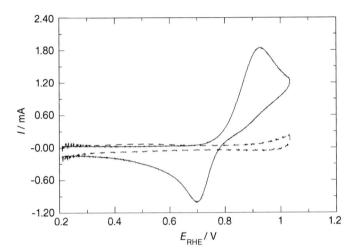

Fig. 3.46 CV of a platinum electrode in an aqueous solution of 2 **mM** *N,N,N′,N′*-tetramethyl-*p*-phenylene diamine in 0.5 **M** H_2SO_4, nitrogen purged; $dE/dt=0.1$ V·s^{-1}; for comparison a CV with supporting electrolyte solution only is shown (–––).

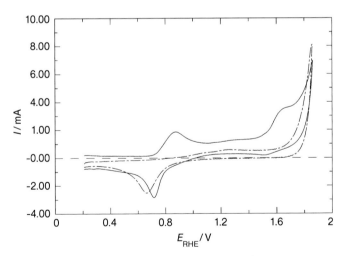

Fig. 3.47 CV of a platinum electrode in an aqueous solution of 2 **mM** *N,N,N′,N′*-tetramethyl-*p*-phenylene diamine in 0.5 **M** H_2SO_4, nitrogen purged, $dE/dt=0.1$ V·s^{-1}; for comparison a CV with supporting electrolyte solution is shown (–––).

Comparison with the CV of the supporting electrolyte solution reveals formation of an oxygen adsorbate on the platinum surface. In the presence of the amine this process is much less pronounced. The peak associated with the reduction of the oxygen adsorbate can be discerned easily from the reduction peak caused by the conversion of the radical cation; confusion is impossible[22].

Literature

R. N. Adams: Electrochemistry at Solid Electrodes, Marcel Dekker, New York 1969.
R. Hand, M. Melicharek, D. I. Scoggin, R. Stotz, A. K. Carpenter, and R. F. Nelson, Collection Czech. Chem. Commun. **36** (1971) 842.

Experiment 3.19: Cyclic Voltammetry of Aromatic Hydrocarbons

Task

With the aid of cyclic voltammetry anodic and cathodic conversions of an aromatic hydrocarbon are studied in a nonaqueous electrolyte solution.

Fundamentals

Both electrooxidation and electroreduction of neutral molecules can result in radical ions. Depending on the composition of the electrolyte solution, these may be consumed in further chemical reactions; they may also be converted into non-radical diions. These in turn may be subject to chemical reactions. Cyclic voltammograms recorded with these molecules will show typical shapes depending on scan rate, potential limits, and other experimental parameters (cf. Fig. 3.44). These CVs provide information needed to estimate or even obtain precisely reaction rates of both chemical and electrochemical reactions. As an example 6,10-diphenylanthracene is considered (Fig. 3.48). Its behavior, in particular the further chemical reaction of the mono- and dication, depend considerably on the water content of the electrolyte solution.

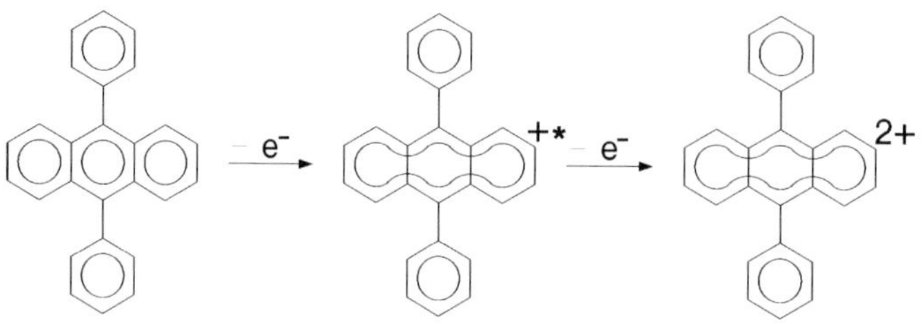

Fig. 3.48 Simplified reaction scheme of transformations of 6,10-diphenylanthracene.

22) This systematic approach is essential; unfortunately recording a background CV is frequently overlooked. This can result in disastrously wrong peak assignments.

Execution

Chemicals and instruments

2 **mM** solution of 6,10-diphenyl anthracene in acetonitrile (purified and dried[23)]) with 0.1 **M** tetraethylammoniumperchlorate (dried)

Purge gas (nitrogen)

2 Platinum electrodes

Silver chloride reference electrode filled with supporting electrolyte solution

Three-electrode cell (H-cell)

Setup for cyclic voltammetry

Setup

The standard setup for cyclic voltammetry (see preceding experiments) is used. The purge gas used to remove oxygen from the electrolyte solution should be piped into a fume hood because of the health risks of acetonitrile.

Procedure

The supporting electrolyte solution is prepared from the solvent and the electrolyte salt. This solution is also used to fill the reference electrode (already containing some pieces of solid silver chloride and the silver wire). In order to avoid contamination with ambient moisture swift working is recommended. The calculated amount necessary to obtain the desired concentration of 6,10-diphenylanthracene ($c_{6,10\text{-diphenylanthracene}} = 2$ **mM**) is added to the main compart-

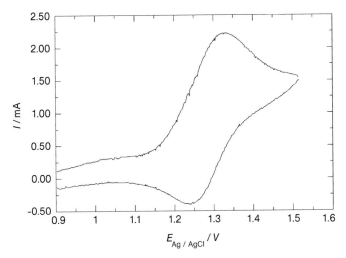

Fig. 3.49 CV of a platinum electrode in a solution of 2 **mM** 6,10-diphenylanthracene + 0.1 **M** tetraethylammoniumperchlorate in acetonitrile, nitrogen purged, $dE/dt = 0.1$ V·s^{-1}.

---

23) Careful drying is essential in this study of the effect of water traces.

ment of the cell. CVs are recorded at various scan rates. Finally measurements are repeated after adding about 2–5 vol% water.

Evaluation

The CV depicted in Fig. 3.49 shows the first redox process. Both subsequent redox processes are visible in Fig. 3.50 recorded with a more positive upper poten-

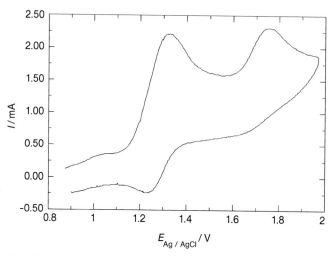

Fig. 3.50 CV of a platinum electrode in a solution of 2 **mM** 6,10-diphenylanthracene + 0.1 **M** tetraethylammoniumperchlorate in acetonitrile, nitrogen purged, $dE/dt = 0.1$ V·s^{-1}.

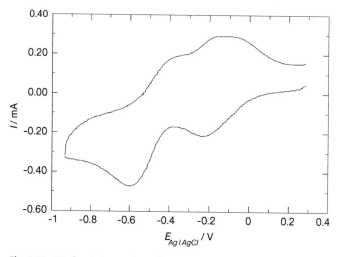

Fig. 3.51 CV of a platinum electrode in a solution of 2 **mM** 6,10-diphenylanthracene + 0.1 **M** tetraethylammoniumperchlorate in acetonitrile, nitrogen purged, $dE/dt = 0.1$ V·s^{-1}.

tial limit. The rapid consumption of the dication by a chemical reaction is evident; there is practically no reduction peak corresponding to the second oxidation peak.

Both reductions are discernible in the CV displayed below (Fig. 3.51).

The fairly rapid chemical reaction of the dianion resulting in an almost complete loss of the corresponding reoxidation peak is obvious.

Upon addition of water, chemical follow-up reactions become so efficient that already the first electron transfer (both reduction and oxidation) becomes irreversible (in the chemical sense of this term).

Literature

R. Dietz and B.E. Larcombe, J. Chem. Soc. B **1970** (1970) 1369.

Experiment 3.20: Cyclic Voltammetry of Aniline and Polyaniline

Tasks

- A polymer film is prepared electrochemically on a metal electrode in a monomer-containing solution
- The polymer film is characterized by electrochemical methods in monomer-free supporting electrolyte solution.

Fundamentals

Most electrochemical synthesis reactions proceed from a substrate molecule to a monomeric product. In some cases formation of poorly defined and mostly undesirable by-products is observed, as in conventional synthetic organic chemistry. These complicate the workup of the product mix and diminish yield. In some cases formation of polymeric products is highly desirable. In electrophoretic coating, electrochemical steps are involved indirectly in film formation only, but in the case of many hetero-atom-containing molecules oxidative formation of a radical cation can result in polymer formation. Radical cations might react with themselves (radical-radical coupling) or with further substrate molecules. Oligomers and finally polymers are formed which will precipitate on the electrode, resulting in a polymer film.

The reaction of aniline – and of its substituted relatives – at a gold or platinum electrode can be studied easily in acidic solution in a three-electrode cell using the experimental setup for cyclic voltammetry. In addition, with this arrangement electrochemically induced changes of the obtained polymer (e.g., changes of color or conductance) can be examined. For an example see Expt. 5.1.

Electropolymerization of aniline at a platinum electrode yielding black products has been known since 1862. In 1910 similar products were obtained by chemical oxidation of aniline, and were called emeraldine and nigraniline. These products are composed of aniline oligomers containing on average eight aniline units coupled via the nitrogen atom and the carbon atom in the para-position of the benzene ring. Electrochemical and spectroscopic investigations have demonstrated the similarity of the chemically and electrochemically prepared products. More recent studies have supported the assumed coupling at the nitrogen and the p-carbon atom. Nevertheless, the true structure, the mechanism of formation, and the effects of chemically or electrochemically induced changes (oxidation and reduction) of polyaniline are still the subject of debate and experiment.

For a few years already compounds of this class, called intrinsically conducting polymers ICP because of the electrical conductance they show upon doping (by, e.g., oxidation) like polyaniline, polypyrrole or polythiophene, have been considered for numerous applications. These range from conductive coatings of packing materials (to prevent electrostatic charging), corrosion protection, replacement of metals in wiring, and molecular electronics to photovoltaics. Accordingly these materials are the subject of intense research.

In an acidic aqueous electrolyte solution containing aniline on a platinum electrode, adsorption and subsequently oxidation of aniline takes place. Further reaction yields polyaniline. This polymer, present as a film on the platinum electrode, shows an interesting electrochemical, electrical, and optical behavior. Whereas the electrical conductance is low in the reduced state it is fairly high in the partially oxidized state, although even higher oxidation results in a poorly conducting state again. Initially colorless to light yellow, the polymer film changes color upon oxidation into emerald green, and finally to black (electrochromism). These changes are reversible, i.e. the properties of the polymer can be switched between the various states.

In this experiment polyaniline is prepared electrochemically on a platinum electrode. Growth of the polymer film is monitored using cyclic voltammetry. In a supporting electrolyte solution oxidation and reduction of the film are studied, and visible changes of the film are interpreted.

Execution

Chemicals and instruments

Aqueous solution of 0.1 **M** aniline + 1 **M** perchloric acid
Aqueous solution of 1 **M** perchloric acid
Purge gas nitrogen
2 Platinum electrodes
Hydrogen reference electrode
Three-electrode cell (H-cell)
Setup for cyclic voltammetry

Setup

The standard setup for cyclic voltammetry (see preceding experiments) is used.

Procedure

In the three-electrode setup with the aniline-containing electrolyte solution, polyaniline is deposited on the platinum working electrode. A platinum counter electrode is used, and the hydrogen reference electrode is filled with perchloric acid only. With a cathodic potential limit $E_{RHE}=0$ V and an anodic potential limit $E_{RHE}=1$ V, CVs are recorded at $dE/dt=0.1$ V·s^{-1}. If no polymer film formation is observed the anodic limit must be increased slightly. Cycles 1 to 10, and the 20th, 30th and 100th cycle are recorded.

Finally the electrode potential is returned to $E_{RHE}=0$ V; the polymer is now in its reduced and almost uncolored state, which is also more stable chemically. The potentiostat is switched into standby mode, the aniline-containing solution is replaced by supporting electrolyte solution, and the film-coated electrode is rinsed carefully with plenty of water and immersed in the cell again.

In the potential range $0.15 < E_{RHE} < 0.9$ V, CVs are recorded at different scan rates ($dE/dt=0.01$; 0.02; 0.03 ... 0.1 V·s^{-1}). Color changes should be monitored.

Overoxidation of the film is studied by stepwise increase of the anodic potential limit in 0.05 V increments between consecutive CV scans up to a value of $E_{RHE}=1.5$ V. After every increase the third CV is recorded.

Instead of aniline, N-methylaniline or o-toluidine can be used at the same concentration. The same potential limits apply; the experimental details are the same.

Evaluation

Figure 3.52 shows a set of CVs recorded during polyaniline deposition.

A reaction equation should be derived going from the monomer to the polymer taking into account electrons and protons. The change of the first anodic peak as a function of time (i.e. cycle number) should be plotted and discussed.

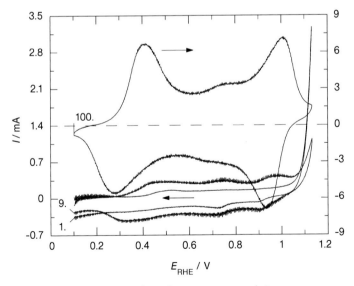

Fig. 3.52 CVs of a platinum electrode in an aqueous solution of 0.1 **M** aniline in 1 **M** HClO$_4$, nitrogen purged, $dE/dt=$ 0.1 V·s^{-1}; cycle numbers indicated.

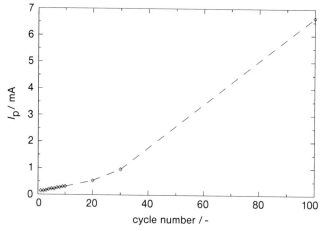

Fig. 3.53 Plot of first peak current in CVs of Fig. 3.52.

The first peak observed in the CVs in Fig. 3.52 is caused by oxidation of the polymer film (transformation of the emeraldine into the nigraniline form); it is not caused by aniline oxidation, which proceeds at much higher electrode potentials also evident in this figure. The amount of deposited polymer corresponds roughly to the height of the current peak. The relationship between cycle number and peak height as displayed in Fig. 3.53 shows noteworthy features. After an increase, initially slow presumably because of the necessary initi-

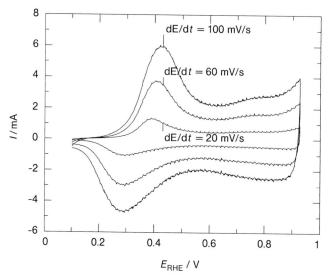

Fig. 3.54 CVs of a polyaniline film on a platinum electrode in an aqueous solution of 1 **M** $HClO_4$, nitrogen purged, scan rates as indicated.

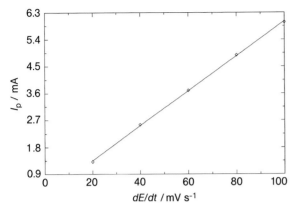

Fig. 3.55 Plot of peak current in CVs of a polyaniline film on a platinum electrode in an aqueous solution of 1 **M** HClO$_4$, as a function of scan rate, nitrogen purged.

alization of film growth by nucleation on the metal surface, the current increases at a faster pace. These results can be employed to estimate the film formation kinetics.

Figure 3.54 shows CVs of the polymer film recorded at different scan rates. The exponent n in the relation $I \approx v^n$ can be derived from a suitable plot as shown in Fig. 3.55.

The exponent in the example shown here is $n=0.94$, which is very close to the value $n=1$ expected for a surface-confined redox system (A. J. Bard and L. R. Faulkner: Electrochemical Methods, Wiley, New York 2001, p. 591).

At higher scan rates mass and charge transport (not transfer at the metal/polymer interface) become dominant; the exponent decreases continuously and approaches the value $n=0.5$ known from cyclic voltammetry.

Electrochromism of the polyaniline film can be discussed qualitatively based on the observed changes as a function of electrode potential, for details see Expt. 5.1.

Successive overoxidation results in CVs are displayed in Fig. 3.56. In the potential range between the two major peaks a new redox peak pair develops, which is assigned to quinoid degradation products. At the most anodic potential limit, this feature dominates the CV.

Literature

K. Menke and S. Roth, Chemie in unserer Zeit **20** (1986) 33.

R. B. Kaner and A. G. MacDiarmid, Scientific American **268(2)** (1988) 106.

P. M. S. Monk, R. J. Mortimer, and D. R. Rosseinsky: Electrochromism: Fundamentals and Applications, VCH, Weinheim 1995.

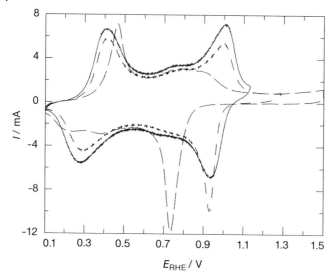

Fig. 3.56 CVs of a polyaniline film on a platinum electrode in an aqueous solution of 1 **M** $HClO_4$ during successive overoxidation.

Experiment 3.21: Galvanostatic Step Measurements [24]

Task

Transition times are measured under galvanostatic conditions.

Fundamentals

A constant current passed through an electrolysis cell results in a reduction reaction at the cathode. In the example studied here the aqueous electrolyte solution contains 1 **M** potassium chloride as supporting electrolyte and 3 **mM** cadmium acetate. The cathodic reaction is

$$Cd^{2+} + 2\ e^- \rightarrow Cd \tag{3.85}$$

A large mercury electrode (mercury pool at the cell bottom) is used as the counter electrode (anode) and at this electrode the potential of a calomel electrode (1 **M**) is established. This electrode maintains a practically constant potential at only small currents, and because of the large surface area correspondingly small current densities are maintained, resulting in negligible potential shifts of the pool electrode. With a small cathode (mercury drop), the situation is that all changes in cell voltage are practically equivalent to changes of cathode potential. To maintain the electrical current at the cathode, cadmium ions can be supplied by diffusion, migration, and convection. Migration is suppressed by the large ex-

---

24) This method is also called chronopotentiometry.

cess of supporting electrolyte salt, resulting in a negligible electric field gradient in the solution (the driving force of migration). Convection is excluded because measurements are performed in a quiet, static solution. The flow of cadmium ions, the electrical current I, the current density j, and the associated concentration gradient can be calculated according to Fick's first law

$$j = z \cdot F \cdot D \cdot \left(\frac{\partial c_{Cd^{2+}}}{\partial x} \right)_{x=0} \tag{3.86}$$

Because a constant current is maintained, the concentration gradients at $x=0$ must have the same slope, as displayed below (Fig. 3.57).

Once the concentration of cadmium ions at the surface reaches zero and reduction of cadmium ions can no longer support the cathodic current, another electrode reaction at a more negative electrode potential will start. In the present example this will be hydrogen evolution by water decomposition. In a plot of cell voltage versus time of measurement, this moment will be indicated by a sudden increase of cell voltage (i.e. cathode potential). The time elapsed from the start of the experiment to this point is called the transition time τ. The relation between current or current density, concentration of cadmium ions, and time at this point can be calculated from $c_{Cd^{2+}}=0$ using Fick's second law:

$$j \cdot \sqrt{\tau} = \left[\frac{z \cdot F \cdot \sqrt{\pi} \cdot \sqrt{D}}{2} \right] \cdot c_{Cd^{2+}} \tag{3.87}$$

This relation is called the Sand equation. At a given concentration the product $j \cdot \sqrt{\tau}$ is constant.

In this experiment a hanging mercury drop is used as the cathode. The major advantage is uncomplicated metal deposition at a clean, well-defined metal sur-

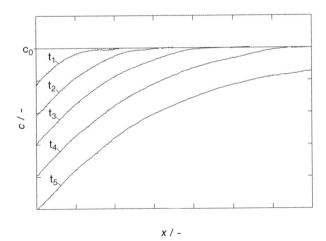

Fig. 3.57 Concentration profiles at various times of measurement $t_1 < t_n < t_5$.

face without significant crystallization overpotential and without formation of metal deposits (because an amalgam is readily formed). This electrode can be prepared fairly easily using a polarograph or a gold wire tip embedded in glass. Strictly speaking, at this drop only the laws of spherical diffusion are valid. At short transition times and/or large drop radii the derived equations will nevertheless transform into the equations provided above. This method can be employed analytically. At various applied currents and a given concentration, transition times are measured and displayed graphically (for an example see below). With a solution containing an unknown concentration of reducible ions the procedure is repeated. At a selected transition time the unknown concentration can be obtained easily according to $I(c_0)/I(c_x)=c_0/c_x$. The practical value of the method is limited by the concentration range yielding transition times which may be measured correctly.

Execution
Chemicals and instruments
Aqueous solution of 0.01 **M** $Cd(CH_3COO)_2$ + 1 **M** KCl
Hanging mercury drop electrode
Mercury
Purge gas (nitrogen)
Beaker or polarograhic cell
Y-t-recorder
Adjustable current source for small current (galvanostat)
Electrical switch

Setup
The electrolyte solution is filled into the cell. Mercury is added, providing the pool electrode at the bottom. The pool is connected to the positive terminal of the current source (by, e.g., a platinum wire embedded in a glass tube immersed in the pool). The drop electrode is connected to the open switch, which is connected to the negative terminal. Both electrodes are connected to the recorder (sensitivity 0.5 V/cm).

Procedure
The solution is purged with nitrogen for about ten minutes. A current of a few microamperes is set. The recorder is started, the switch is closed. After observing the transition time the switch is opened, the nitrogen purge is repeated in order to remove any concentration gradient. The procedure is repeated with different current settings. Suitable combinations resulting in measurable transition times must be determined empirically.

Evaluation
A typical set of voltage vs. time curves is shown in Fig. 3.58. A plot of currents as a function of transition times yields Fig. 3.59.

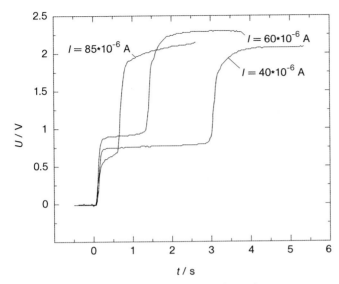

Fig. 3.58 Typical voltage vs. time plots obtained in a chrono-
potentiometric experiment.

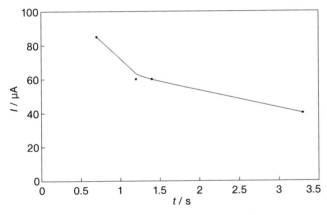

Fig. 3.59 Plot of current vs. transition time with data displayed in Fig. 3.58.

Questions
- Why is it possible to use equations applicable for linear diffusion instead of those for spherical diffusion?
- Is this experiment also possible without supporting electrolyte or in stirred solution?

Literature
A. J. Bard and L. R. Faulkner: Electrochemical Methods, Wiley, New York 2001, p. 305.

Experiment 3.22: Chronoamperometry [25]

Task

The surface area of an electrode is determined via a chronoamperometric measurement.

Fundamentals

A sudden potential step applied within a very short time (ideally instantaneous) to an electrode in contact with an electrolyte solution containing a dissolved redox system from an initial to a final value results in a flow of electrical current. It is caused by the change of electrolyte composition at the interface (in particular the concentration of both species of the redox system) required according to the Nernst equation by the new potential value. Initially the current is very large, being essentially limited only by the Ohmic resistance of the electrolyte solution. Part of the current is consumed for recharging the electrochemical double layer. Because mass transfer from the bulk of solution rapidly becomes the current-limiting factor the current will soon decrease. At a surface concentration zero of the consumed species, only the diffusion-limited current $I_{\text{lim,diff}}$ will flow; this current is controlled by the concentration profile of the consumed species. As a function of time this profile extends deeper into the solution its slope decreases accordingly and so does the current supported by the diffusion flow. As a result of the mathematical treatment, the Cottrell equation is obtained (F.G. Cottrell, Z. Phys. Chem. **42** (1902) 385):

$$I(t) = I_{\text{D}}(t) = \frac{n \cdot F \cdot \sqrt{D} \cdot c_0 \cdot A}{\sqrt{\pi} \cdot \sqrt{t}} = k_{\text{Cot}} t^{-1/2} \tag{3.88}$$

From the constant product $I \cdot t^{1/2}$ various experimental parameters can be derived.

Execution

Chemicals and instruments

Aqueous solution of 0.005 **M** $K_4Fe(CN)_6$ + 0.5 **M** K_2SO_4

2 Gold electrodes

Mercurous sulfate reference electrode

Purge gas nitrogen

Potentiostat

H-cell

Y-t-recorder or transient recorder or computer with sufficiently fast ADDA-converter and software

---

[25] This method is also called potentiostatic step method because of the applied potential vs. time program.

Setup

The gold electrodes and the reference electrodes are placed in the cell filled with electrolyte solution. The potentiostat is connected with the electrodes and the computer (in the example described here).

Procedure

After purging the electrolyte solution with nitrogen, potential steps from the spontaneously established rest potential (and may need some time to reach a stationary value because the second component of the redox system is initially missing). Current transients are recorded after potential steps to values where the current is diffusion limited.

Evaluation

Figure 3.60 shows a typical transient recorded after a potential step from $E_{i,Hg_2SO_4} = -0.63$ V to $E_{f,Hg_2SO_4} = 0.0$ V.

From the obtained data the surface area of the employed spherical gold electrode could determined as $A = 0.41$ cm^2.

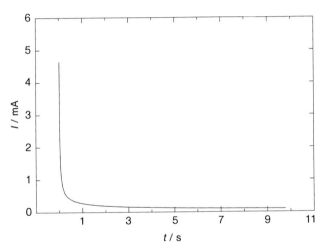

Fig. 3.60 Current transient after a potential step from $E_i = -0.63$ V to $E_f = 0.0$ V, gold electrode, aqueous solution of 0.005 M K$_4$Fe(CN)$_6$ + 0.5 M K$_2$SO$_4$, nitrogen purged.

Experiment 3.23: Chronocoulometry

Task
The surface of an electrode is determined via a chronocoulometric measurement.

Fundamentals
With the potential of a working electrode set to a value where the electrode reaction proceeds at a rate limited only by diffusion, the charge consumed by the reaction can be described by the integrated form of the Cottrell equation:

$$Q = \frac{2 \cdot n \cdot F \cdot \sqrt{D} \cdot c_0 \cdot A \cdot \sqrt{t}}{\sqrt{\pi}} \tag{3.89}$$

In addition to this charge, the charge Q_{DL} needed for charging the double layer from the value corresponding to the initial electrode potential to the potential set by the step and the charge needed for conversion of those species already present on the electrode at the time of the step (by, e.g., adsorption), Q_0, have to be considered. Since both Q_{DL} and Q_0 do not depend on time hey can be easily obtained by extrapolation to $t=0$ s, and the resulting value can be subsequently subtracted. According to Eq. (3.89) chronocoulometry can be applied to determine concentrations, diffusion coefficients, or surface areas.

At first glance this method seems to be equivalent to chronoamperometry (see Expt. 3.22). Because in addition a special electronic device (an integrator or coulometer) not necessarily present in every laboratory is needed, this method seems to be fairly unattractive. Considering the central fact that charge – the system's response signal towards the potential step – is growing during the experiment and only charge data far away from the potential step are evaluated, and these are most likely not affected by any transient behavior of the setup (any distortions by ringing effects etc.), the method looks attractive indeed. In addition the integration has a smoothing effect, i.e. noise on the current signal will be averaged[26]. Both advantages are not effective in chronoamperometry, and in addition the determination of Q_{DL} and Q_0 is not possible.

Execution
Chemicals and instruments
Aqueous solution of 0.005 M $K_4Fe(CN)_6$ + 0.5 M K_2SO_4
2 Gold electrodes
Mercurous sulfate reference electrode
Purge gas nitrogen
Potentiostat
H-cell

---

26) Instead of an analog integrator a fast AD converter card can be used. The charge can be obtained easily by numerical integration. The smoothing effect is less important.

Analog integrator and Y-t-recorder or transient recorder or computer with sufficiently fast ADDA-converter and software

Setup
The gold electrodes and the reference electrodes are placed in the H-cell filled
with electrolyte solution. The potentiostat is connected with the electrodes and
the computer (in the example described here).

Procedure
After purging the electrolyte, the potential step is applied, and the consumed
charge is registered.

Evaluation
A typical Q-t-plot is shown in Fig. 3.61. The contributions from Q_{DL} and Q_0 are
negligible; the line passes almost exactly through the origin. The surface area of
the gold electrode employed in this experiment is $A=0.47$ cm^2.

Literature
F.C Anson, Anal. Chem. **38** (1966) 54.

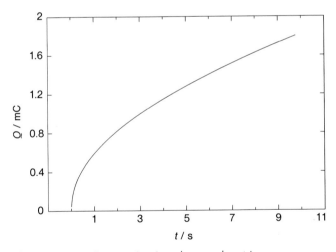

Fig. 3.61 Current-time transient in a chronocoulometric experiment with a gold electrode, aqueous solution of 0.005 **M**
K$_4$Fe(CN)$_6$ + 0.5 **M** K$_2$SO$_4$, nitrogen purged, potential step
from $E_{MSE}=-$ 0.67 V to $E_{MSE}=0.0$ V.

Experiment 3.24: Rotating Disc Electrode

Tasks
- The diffusion coefficient of the ion $Fe(CN)_6^{3-}$ is determined.
- The exchange current density of the redox $Fe(CN)_6^{3-/4-}$ system is determined.

Fundamentals
In many electrode processes the flow of electric current is not limited by the charge transfer step itself but by relatively slower steps like, e.g., transport or chemical reaction. If a mathematical treatment is available enabling the calculation of the influence of, e.g., mass transport and the subsequent elimination of this influence from the experimental data, the charge transfer itself may be studied and the current-potential plot of the charge transfer reaction may be obtained. There are only a few systems where this procedure was successfully applied; one of these examples is the rotating disc electrode. In this setup the electrode is a circular disc embedded in the smooth axial surface of a cylindrical body made of insulating material (Fig. 3.62). In addition, a ring-shaped electrode may be placed around the disc enabling products formed at the disc to be studied (see following experiment).

This experiment explores various experimental possibilities of the rotating disc electrode RDE.

Execution
Chemicals and instruments
Aqueous stock solution of 0.1 M $K_3Fe(CN)_6$ + 0.5 M K_2SO_4
Aqueous stock solution of 0.1 M $K_4Fe(CN)_6$ + 0.5 M K_2SO_4
Aqueous solution of 0.05 M H_2SO_4
Aqueous stock solution of 0.1 M $Fe(NH_4)(SO_4)_2 \cdot 12\ H_2O$

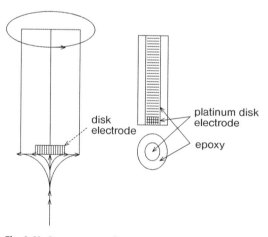

Fig. 3.62 Cross section of a rotating disc electrode and flow profile.

Aqueous stock solution of 0.1 **M** $FeSO_4 \cdot 7\ H_2O$
Cell with rotating platinum disc electrode and controller
Platinum wire counter and reference electrode
Potentiostat and function generator or computer with sufficiently fast ADDA-
converter and software
X-Y-recorder
Pipette 5 ml
Pipette 10 ml
Purge gas (nitrogen)

Setup
The rotating disc electrode is connected to the speed control/power supply unit,
and the disc electrode, the counter, and the reference electrodes are connected
to the potentiostat. In the example described below a disc electrode with
$A=0.29\ cm^2$ was used, and a silver/silver chloride system was used as a reference electrode. Instead a platinum wire may be used; at the wire the redox potential defined by the concentrations (activities) of the redox components will be
established. The solution is purged with inert gas for about 20 min, and the gas
supply is reconnected to an inlet providing a gas blanket above the solution.

Procedure
After purging the electrolyte a CV of the redox system in stagnant (unstirre) solution containing 5 **mM** $K_3Fe(CN)_6 + 5$ **mM** $K_4Fe(CN)_6 + 0.5$ **M** K_2SO_4 at $dE/$
$dt = 0.1\ V \cdot s^{-1}$ of the reaction

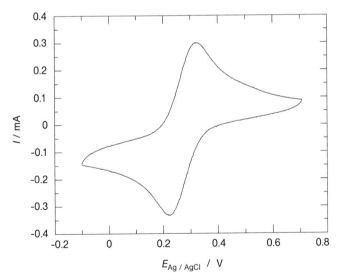

Fig. 3.63 CVs of a platinum electrode in an aqueous solution
5 mM $K_3Fe(CN)_6 + 5$ mM $K_4Fe(CN)_6 + 0.5$ M K_2SO_4 at
$dE/dt = 0.1\ V \cdot s^{-1}$.

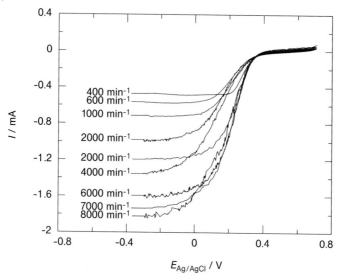

Fig. 3.64 CVs (only negative-going part shown, no smoothing of data) of a platinum electrode in an aqueous solution of 5 mM $K_3Fe(CN)_6$ + 5 mM $K_4Fe(CN)_6$ + 0.5 M K_2SO_4 at $dE/dt = 0.01$ V·s^{-1}; angular velocities as indicated.

$$Fe(CN)_6^{3-} + e^- \rightleftarrows Fe(CN)_6^{4-} \tag{3.90}$$

is recorded. A CV as displayed in Fig. 3.63 is obtained (see also Expt. 3.14).

1. In the first part of the experiment, current density vs. electrode potential curves are recorded at various rates of rotation to examine the expected increase of the diffusion-limited current as a function of the rate of rotation. A typical set of results is displayed below in Fig. 3.64; suggested rates of rotation are 200 ... 800 min^{-1}.

2. The slope of the plot of the limiting current density j_{lim} vs. the square root of the rate of rotation $\omega^{1/2}$ yields the diffusion coefficient of the trivalent complex cation. If a film of Prussian blue (not necessarily visible to the eye, but causing distortions in the CV) is formed it can be removed by dipping the working electrode alternately into sulfuric acid and an aqueous solution of ammonia.

In the second part of the experiment, current density vs. electrode potential curves are recorded at various rates of rotation and at very low overpotentials, i.e. far away from the diffusion-limited case. The results yield data necessary to obtain Tafel plots which in turn provide the exchange current density j_0.

Evaluation
The theory of the rotating disc electrode yields a relationship between rate of rotation or angular velocity and limiting current

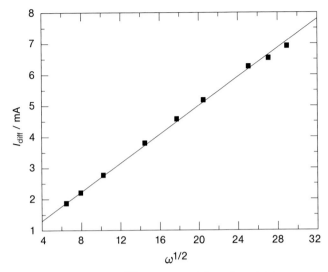

Fig. 3.65 Plot of I_{diff} vs $\omega^{1/2}$ (experimental data of Fig. 3.64).

$$I_{diff} = 0.62 \cdot n \cdot F \cdot A \cdot D^{2/3} \cdot v^{-1/6} \cdot c \cdot \omega^{1/2} \tag{3.91}$$

with: I_{diff} = diffusion-limited current in A

$\quad\quad n$ = number of electrons transferred in the reaction (1)

$\quad\quad D$ = diffusion coefficient of consumed species in $cm^2 \cdot s^{-1}$

$\quad\quad v$ = kinematic viscosity of the electrolyte solution (here: $1 \cdot 10^{-2}\ cm^2 \cdot s^{-1}$)

$\quad\quad c$ = concentration of consumed species (here in $mol \cdot cm^{-3}$)

$\quad\quad v$ = angular velocity $\omega = 2 \cdot \pi \cdot f$ in s^{-1}

$\quad\quad f$ = frequency (rate) of rotation in s^{-1}

From

$$\dot{j}_{diff} = \frac{I_{diff}}{A} \tag{3.92}$$

and the slope of the plot, $\tan a\,(A \cdot cm^{-2} \cdot s^{1/2})$, the diffusion coefficient can be calculated according to

$$D = \left(\frac{\tan a \cdot v^{1/6}}{0.62 \cdot n \cdot F \cdot c} \right)^{3/2} \Bigg/ cm^2 \cdot s^{-1} \tag{3.93}$$

The results shown yield $D = 0.66 \cdot 10^{-5}\ cm^2 \cdot s^{-1}$. Literature values for $Fe(CN)_6^{3-}$ are $D = 0.66 \cdot 10^{-5}\ cm^2 \cdot s^{-1}$, for $Fe(CN)_6^{4-}$ $D = 0.57 \cdot 10^{-5}\ cm^2 \cdot s^{-1}$ (K. J. Kretschmar, C. H. Hamann and F. Faßbender, J. Electroanal. Chem., **60** (1975) 239) and for $Fe(CN)_6^{3-}$ are $D = 1.18 \cdot 10^{-5}\ cm^2 \cdot s^{-1}$ (Handbook of Chemistry and Physics 71st Ed., 1991, p. 6–151).

To obtain Tafel plots (i.e. a semi-logarithmic display of current vs. overpotential) determined only by charge transfer from the CVs obtained with the RDE at

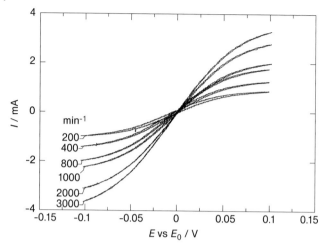

Fig. 3.66 CVs of a rotating platinum electrode in an aqueous solution of 5 mM $K_3Fe(CN)_6$ + 0.5 M K_2SO_4 at $dE/dt =$ 5 mV·s^{-1}; angular velocities as indicated; platinum wire reference electrode.

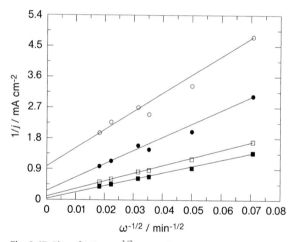

Fig. 3.67 Plot of $1/j$ vs $\omega^{-1/2}$ extrapolation towards $\omega \to \infty$.

various angular velocities, current/potential values are taken, plotted, and extrapolated to infinite angular velocity (EC:193). Figure 3.66 shows the raw data, Fig. 3.67 the next step.

A Tafel plot of the charge transfer currents (no longer influenced by mass transport) obtained by extrapolation in Fig. 3.67 is shown below.

From the Y-axis intercept the exchange current density j_0 can be obtained and from the slope the symmetry coefficient a and the number of electrons n transferred in the charge transfer step. In this example $n = 1$, $j_0 = 2.1$ mA·cm^{-2}.

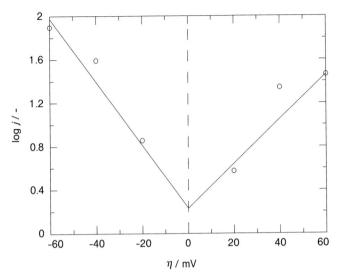

Fig. 3.68 Tafel plot of the data from Fig. 3.68.

Question
Which overpotential is influenced in a predictable way by rotating the electrode?

Literature
W. J. Albery and M. L. Hitchman: Ring-Disc Electrodes, Clarendon Press, Oxford 1971.
B. Gostisa-Mihelcic and W. Vielstich, Ber. Bunsenges. Phys. Chem. **77** (1973) 476.
A. J. Bard and L. R. Faulkner: Electrochemical Methods, Wiley, New York 2001.
Yu. V. Pleskov and V. Yu. Filinovskii: The Rotating Disc Electrode, Consultants Bureau, New York 1976.

Experiment 3.25: Rotating Ring-Disc Electrode

Tasks
- The collection efficiency N of a rotating ring-disc electrode is determined.
- The mechanism of the electroreduction of copper(II)-ions is investigated.

Fundamentals
The products of electrode processes at the rotating disc electrode are transported away sideways with the flux of species. Their further investigation both qualitatively (i.e. their identification) as well quantitatively is possible with a ring electrode embedded in the electrode body closely surrounding the disc electrode with only a thin, insulator-filled gap between them. This arrangement shown schematically in Fig. 3.69 is called a rotating ring-disc electrode RRDE.

Identification of species generated at the disc electrode is possible by further electrochemical conversion at the ring electrode set at a suitable electrode poten-

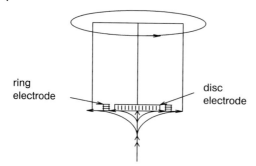

Fig. 3.69 Cross section of a rotating ring-disc electrode and flow profile.

tial. Because transport of particles from the disc to the ring can be calculated precisely, the measurement of the ring provides in addition quantitative information. If the species generated at the ring are subject to a competing homogeneous chemical reaction, the amount of species actually detected at the ring (which is lower than expected without the chemical reaction) yields information regarding the rate of the homogeneous chemical reaction.

The diffusion-limited current at a ring electrode can be calculated using the mathematics already employed with the disc electrode:

$$I_{R,diff} = 0.62 \cdot n \cdot F \cdot \pi \cdot \left(r_3^2 - r_2^2\right)^{2/3} \cdot D^{2/3} \cdot \sqrt{\omega} \cdot v^{-1/6} \cdot c \tag{3.94}$$

The ratio of the ring current to the disc current for a given electrode reaction thus depends only on the radii:

$$\frac{I_{R,diff}}{I_{D,diff}} = \frac{\left(r_3^2 - r_2^2\right)^{2/3}}{r_1^2} \tag{3.95}$$

The possibility to detect species at the ring formed during a reaction at the disc as intermediates is particularly attractive in attempts to elucidate reaction mechanisms. The fraction of species generated at the disc and detected at the ring is given by N, which is independent of the angular velocity. Because of the complicated mathematics, values of N are tabulated for sets of ratios of the radii r_3/r_2 and r_2/r_1 (see literature). The electrode employed here showed $r_1 = 2.28$ mm, $r_2 = 2.58$ mm, and $r_3 = 2.73$ mm. An approximate calculation of N is possible (V. Yu. Filinovsky and Yu. V. Pleskov in: Comprehensive Treatise of Electrochemistry Vol. 9 (E. Yeager, J.O'M. Bockris, B. E. Conway and S. Sarangapani, Eds.) Plenum Press, New York 1984, p. 339):

$$N = \left(\frac{r_3^3 - r_2^3}{r_2^3 - r_1^3}\right)^{2/3} \left[\frac{1}{2.44 + \left(r_1^3/\left(r_2^3 - r_1^3\right)\right)^{2/3}} + \frac{1}{2.44 + \left(\left(r_3^3 - r_2^3\right)/\left(r_2^3 - r_1^3\right)\right)^{2/3}} \right.$$

$$\left. - \frac{1}{2.44 + \left(\left(r_1^3/r_3^3\right) \cdot \left(\left(r_3^3 - r_2^3\right)/\left(r_2^3 - r_1^3\right)\right)\right)^{2/3}} \right] \tag{3.96}$$

This yields a theoretical value of $N=0.154$ for the given dimensions.

The electroreduction of copper(II)-ions is studied as a typical example of a reaction where identification of an intermediate is pivotal in elucidating the mechanism. In this reaction two pathways are conceivable: direct reduction via a simultaneous transfer of two electrons or initial reduction yielding a copper(I)-ion which may be reduced further subsequently:

$$Cu^{2+} + e^- \rightarrow Cu^+ \tag{3.97}$$

$$Cu^+ + e^- \rightarrow Cu \tag{3.98}$$

Distinction between the mechanisms depends on the proof of existence of copper(I)-ions. This could be possible by keeping the ring electrode at a potential where copper(I)-ions are reoxidized. Further competing reactions must be definitely excluded. If a ring current is detected, the second mechanism is proven; if no ring current, the direct reduction (first mechanism).

Further kinetic data may be obtained with a ring-disc electrode. If the product formed at the ring is consumed by a homogeneous chemical reaction the actually observed (apparent) value of N will be diminished. From measurements of j_R at a suitably set electrode potential the rate of reaction of the homogeneous reaction can be elucidated.

Execution
Chemicals and instruments
Aqueous solution of **1 mM** $CuCl_2$ **+ 0.5 M KCl**
Cell with rotating platinum ring-disc electrode and controller
Platinum wire counter electrode
Reference electrode (saturated calomel, silver/silver chloride)
Bipotentiostat, function generator, and two X-Y-recorders or computer with ADDA-converter and software
Purge gas (nitrogen)

Setup
The rotating ring-disc electrode is connected to the speed control/power supply unit, and the ring, disc, counter, and reference electrode are connected to the bipotentiostat. In the example described below a mercurous sulfate reference electrode (E_{MSE}) was used. The solution is purged with inert gas for about 20 min, and the gas supply is reconnected to an inlet providing a gas blanket above the solution. This is particularly important because at high angular velocities the electrolyte solution may form a vortex with an extended surface area enhancing solution/gas contact. Traces of residual dioxygen might be dissolved and subsequently reduced, thus simulating the presence of copper(I)-ions.

Procedure
Figure 3.70 shows a typical set of results at different angular velocities and at a fixed ring potential. At moderately negative (cathodic) electrode potentials of the disc electrode the reaction stops at the copper(I)-ion (Eq. 3.97); these ions are re-

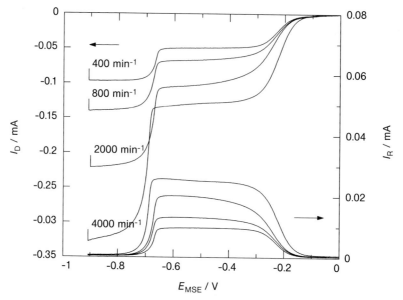

Fig. 3.70 Rotating disc and ring electrode currents (platinum electrodes) in an aqueous solution of 1 **mM** CuCl$_2$ + 0.5 **M** KCl$_4$ at $dE/dt = 10$ mV·s^{-1}; angular velocities as indicated; $E_{R,MSE} = 0.2$ V.

oxidized at the ring electrode. At more negative disc electrode potentials the reduction proceeds straight to copper(0), the ring current collapses. From ring and disc currents a value of $N = 0.19$ is calculated.

Literature

W. J. Albery and S. Bruckenstein, Trans. Faraday Soc. **62** (1966) 1920.

W. J. Albery and M. L. Hitchman: Ring-Disc Electrodes, Clarendon Press, Oxford 1971.

Yu. V. Pleskov and V. Yu. Filinovskii: The Rotating Disc Electrode, Consultants Bureau, New York 1976.

Experiment 3.26: Measurement of Electrode Impedances

Task

The impedance of a platinum electrode in contact with an electrolyte solution containing a redox system is measured, and the exchange current density of the redox reaction is determined.

Fundamentals

In a potentiostatic three-electrode arrangement the potential of the working electrode can be modulated (stimulated, disturbed) by a variety of different types of signals (steps, sweeps, square waves etc.). A particularly powerful modulation employs a sine wave of small amplitude in a frequency range from a few milli-hertz to a megahertz. The evaluation of the phase and amplitude relation be-tween the modulating voltage and the systems response (cell current), i.e. the impedance, provides access to a wealth of information including structural and kinetic data. The small amplitude allows various approximations in the mathe-matical treatment of the various steps of the electrode reaction.

In the most basic case of a redox electrode, this reaction is composed of trans-port of the reacting species, charge transfer, and finally transport away of the re-action product. At high frequencies of the modulating sine wave (a few kilohertz) the contributions of transport can be neglected. Taking into account diffusion however – and the numerous available software packages provide easy possibilities – the equivalent circuit suggested first by Randles can be used (Fig. 3.71):

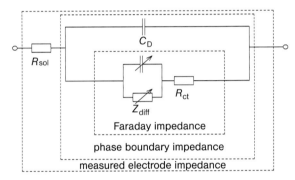

Fig. 3.71 Equivalent circuit according to Randles.

In addition to the use of equivalent circuits taken from electrical engineering, other approaches based on transfer functions have been proposed. These are less instructive and obvious in the example studied here, and their application would not provide additional insights.

Execution
Chemicals and instruments
Aqueous solution of 0.01 **M** $(Fe(NH_4)_2(SO_4)_2 + Fe(NH_4)(SO_4)_2 + 1$ **M** $HClO_4$
Cell for AC measurements (see chapter 1)
Platinum wire counter and reference electrode
Platinum sphere working electrode
Purge gas (nitrogen)
Impedance measurement setup[27]

---

27) Because of the large number of commercially available setups
including software for both measurement and evaluation, in most
cases no details of a particular instrument are discussed here.

Setup

Impedance measurement setup, potentiostat, electrodes, and cell are connected. The redox system dissolved in the electrolyte solution provides the formal potential E_0 at the platinum wire reference electrode; this type of reference electrode in addition avoids contamination of the solution possibly caused by constituents of other reference electrodes.

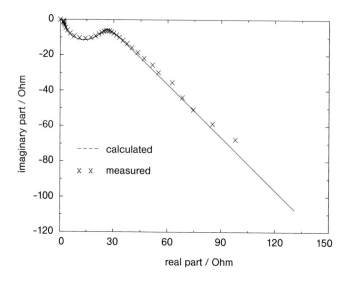

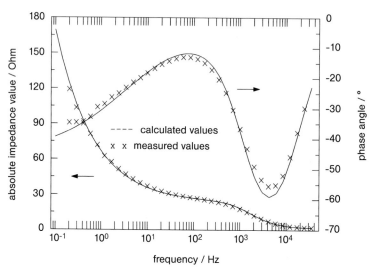

Fig. 3.72 Impedance of a platinum electrode in contact with an aqueous solution of 0.01 **M** $(Fe(NH_4)_2(SO_4)_2 + Fe(NH_4)(SO_4)_2 + 1$ **M** $HClO_4$; $E = E_0$; top: complex plane plot; bottom: Bode plot.

Procedure

The solution is rigorously purged with inert gas. The suitable state of the surface of the working electrode is verified by recording a CV in the electrolyte solution. If it does not agree with literature data, the working electrode should be cleaned carefully; subsequently it should be exposed to electrode potential cycles in perchloric acid between (without the redox system being present in solution) hydrogen and oxygen evolution onset until a CV again in agreement with literature data is obtained.

The impedance measurement is performed with the electrolyte solution containing the redox system in the frequency range 1 Hz $< f <$ 100 kHz (the upper limit may be slightly lower depending on the instrument actually used). Evaluation can be based on the simple equivalent circuit depicted in Fig. 3.72.

Evaluation

A typical result displayed in the complex plane plot is shown in Fig. 3.72 top; the Bode plot is shown in Fig. 3.72 bottom. From the displayed results a charge transfer resistance $R_{ct} = 21.75\ \Omega$ and a double-layer capacity $C_D = 58\ \mu F$ were calculated. The exchange current density is $j_0 = 3.94\ A \cdot cm^{-2}$. Taking into account the redox component concentrations, the standard exchange current density is $j_{00} = 0.394\ A \cdot cm^{-2}$.

Experiment 3.27: Corrosion Cells

Tasks

- Cell voltages and short-circuit currents of various corrosion cells are measured.
- The function of a magnesium electrode as a sacrificial anode is examined.

Fundamentals

Two metals of different nobility (i.e. at different positions in the electrochemical series) electrically connected form a corrosion element when brought into contact with an electrolyte solution: the less noble metal will be dissolved anodically whereas at the more noble metal either dioxygen reduction or hydrogen evolution (depending on the solution composition and the identity of the metal) will proceed. Such corrosion elements may be formed on a microscopic scale by deposition of tiny metal particles on foreign metal surfaces; they can also be formed on a macroscopic scale by, e.g., using screws, nuts, and washers of different metals in construction work. Because of the size of the electrodes employed here we will call the system a macro corrosion element. It enables the instructive and easily understood investigation of technologically relevant corrosion processes (see also contact corrosion, local cell formation).

Measurement of the electrical voltage U_0 established between the pieces of metal (i.e. of the electrode potential difference) permits the easy identification of the more noble metal (the cathode) and the less noble, actually corroding, metal (the anode). The short-circuit current I_{sc} measurable between the metals indicates the possible rate of corrosion. These measurements are possible with very simple instruments: a voltmeter and an ampere meter[28].

Execution
Chemicals and instruments
Aqueous solution of NaCl 3.5wt%
Pieces of aluminum, copper, iron, magnesium, and zinc sheets, preferably of similar size
Beaker
2 Multimeters
pH Indicator paper
Abrasive paper
Ethanol

Setup
The metal pieces to be investigated are connected with the voltage input of the multimeter and inserted into the sodium chloride solution. For measurement of I_{sc} the multimeter is set to current measurement.

In the investigation of the function of the magnesium sacrificial anode the multimeters are set to current measurement and connected in series between the copper and the iron electrode. The magnesium electrode is connected with the connector between the multimeters.

Procedure
The metal electrodes are cleaned with abrasive paper in order to remove oxides and other surface layers. They are degreased with ethanol and after drying immersed in the electrolyte solution. When a constant value of U_0 is observed it is recorded. Measurements of I_{sc} are performed in the same manner, a very constant value may be impossible to observe.

Evaluation
Table 3.3 shows the investigated combinations of metals and typical values of voltages and currents. The polarities found with these macro corrosion elements can be explained easily based on the electrochemical series. The short-circuit currents can be understood qualitatively also based on the series. Further evaluation of I_{sc} is unsafe because of slight differences in size of the metal pieces employed here and because of other experimental factors (e.g., distance between electrodes).

---

28) Because of the finite input impedance of an ampere meter ($R_{in} > 0\ \Omega$) this short-circuit case can only be approximated. With a sufficiently sensitive instrument employ- ing a low value current shunt resistor the error is negligible. A current follower with an operational amplifier avoids this error completely.

Table 3.3 Results of corrosion measurements [29].

Metals	U_0/mV	Polarity	I_{sc}/mA
Fe–Al	480	Fe(+); Al(–)	1
Fe–Zn	800	Fe(+); Zn(–)	5
Fe–Mg	1160	Fe(+); Mg(–)	–
Zn–Al	288	Al(+); Zn(–)	0.1
Cu–Zn	180	Cu(+); Zn(–)	very small
Cu–Al	130	Cu(+); Al(–)	very small
Cu–Mg	1370	Cu(+); Mg(–)	–
Cu–Fe	334	Cu(+); Fe(–)	0.2
		$I_{sc,Cu-Fe}$/mA	$I_{sc,Mg-Fe}$/mA
Cu–Fe	–	–	–
Cu–(Mg)Fe	–	–	7.7

The function of the magnesium electrode as a sacrificial anode can be understood quite easily, and the value of I_{sc} is particularly large according to the position of this metal in the electrochemical series. With the indicator paper a pH shift at the iron electrode towards alkaline values can be demonstrated, this indicates hat dioxygen reduction is proceeding at the iron electrode yielding hydroxyl ions as a reaction product.

Questions

Why does I_{sc} drop as a function of time?

Literature

Uhlig's corrosion handbook (R.W. Revie, H.H. Uhlig, Eds.), Wiley, New York 2000.
P. R. Roberge: Corrosion Basics: An Introduction (2nd ed.), NACE International, Houston, Texas, USA 2006.
H. Kaesche: Corrosion of metals, Springer, Berlin 2003.

Experiment 3.28: Aeration Cell

Task

In an aeration cell with two iron nails as electrodes the influence of local differences in dioxygen concentration on corrosion is studied.

Fundamentals

The cathodic reaction in most corrosion processes is the electroreduction of dioxygen; only in acidic environments are proton reduction and hydrogen evolu-

---

29) Observed values may differ considerably because oxide or hydroxide layers formed on the metal pieces exposed to air and moisture may interfere with establishment of an electrode potential.

tion a viable possibility. Local differences in dioxygen concentration may induce gradients of electrode potential and accordingly locally different electrode reactions. At places of high dioxygen concentration its electroreduction prevails, whereas at places of low concentration anodic metal dissolution proceeds. Because a high local dioxygen concentration is supported by a steady supply of air these corrosion elements are called aeration cells (sometimes also differential aeration cells). Corrosion involving dioxygen reduction is by far the most common form of corrosion of metals; the actual damage to an economy may amount to several percent of gross national product.

Execution
Chemicals and instruments
Two large iron nails
Aqueous solution of NaCl 3.5wt%
Beaker
Ampere meter
Voltmeter
Air pump
Glass tube
Glass tube closed at one end with a porous frit or other porous material
Abrasive paper
Ethanol

Setup
The sodium chloride solution is filled into the beaker, and the glass tube is immersed in the solution down to the bottom and fixed to the wall of the beaker. The tube with the frit is connected to the air pump and also immersed in the beaker with its bottom slightly higher than the lower end of the open tube. The nails are cleaned with abrasive paper in order to remove oxides and other surface layers. They are degreased with ethanol and, after drying, immersed in the electrolyte solution, one into the tube, and one into the main volume in the beaker. The nails are connected to the voltmeter.

Procedure
When a constant value of U_0, around 0 V (why?), is observed, the air pump is switched on. After a few minutes a new value of U_0 is observed distinctly different from the initial one. Now I_{sc} is measured. The pump is switched off, and I_{sc} is observed as a function of time.

Evaluation
From the observed value of U_0 and its polarity the corroding electrode (the anode) can be identified; this can be supported by observing visible changes (traces of corrosive attack and metal dissolution) at the anode. In a typical experiment a value of $U_0 = 60$ mV was recorded, $I_{sc} = 2$ mA. After switching off the air supply the value dropped rapidly to $I_{sc} = 0.5$ mA.

Literature
Uhlig's corrosion handbook (R.W. Revie, H.H. Uhlig, Eds.), Wiley, New York 2000.
P.R. Roberge: Corrosion Basics: An Introduction (2nd ed.), NACE International, Houston, Texas, USA 2006.
H. Kaesche: Corrosion of metals, Springer, Berlin 2003.

Experiment 3.29: Concentration Cell

Task
The cell voltage of a concentration cell composed of two identical metal electrodes immersed in solutions of different concentrations of the respective metal ions is measured. The corrosion current caused by this potential difference is measured.

Fundamentals
According to the Nernst equation (EC:81), at a metal immersed into a solution containing ions of this metal an electrode potential is established. Its value depends on the concentration (or more precisely: the activity) of the ions. When large pieces of metal with large surface areas are immersed, local metal ion concentrations may differ. The part of the metal immersed at a place of higher concentration will form a cathode, and the part in contact with more dilute solution will be the anode. The electrical current carried by electrons in the metal between these regions is associated with an ionic current in solution supported by metal dissolution at the anodic part and metal deposition at the cathodic one. The electrochemical system thus established is called a concentration cell (EC:86; 105)

Execution
Chemicals and instruments
Two copper electrodes
Aqueous solution of $CuSO_4$ 1 **M**
Aqueous solution of $CuSO_4$ 0.01 **M**
Beaker
Ampere meter
Voltmeter
Glass tube closed at one end with a porous frit or other porous material
Abrasive paper
Ethanol

Setup
One solution is poured into the beaker. The glass tube is immersed into the solution and filled with the second solution. The glass frit (or some other porous material attached to the glass tube) prevents mixing of the solutions. The cleaned and degreased copper electrodes are inserted into the solutions.

Procedure

The voltage between both copper electrodes is measured with the voltmeter, and the short-circuit current is measured with the ampere meter once a stable voltage has been observed.

Evaluation

In a typical experiment $U_0 = 36$ mV is found, and the copper electrode in the concentrated solution is the anode, as expected. Deviations of the observed values from the calculated ones are caused by the activity coefficients of the copper ions which are considerably different from unity. A short-circuit current $I_{sc} = 0.25$ mA was measured.

Literature

H. Kaesche: Corrosion of metals, Springer, Berlin 2003.

Experiment 3.30: Salt Water Drop Experiment According to Evans

Tasks

- The local distribution of corrosion reactions on a steel plate is verified by the salt water drop experiment according to Evans.
- The formation of local corrosion elements caused by mechanical stress at an iron nail is observed.

Fundamentals

Corrosion, in particular the anodic and cathodic reactions, and its local distribution can be verified easily by detection of corrosion products. During corrosion of iron-based alloys the dissolution of iron constitutes the anodic reaction. The initially form Fe(II) ions can be detected with sodium hexacyanoferrate(III) by the formation of strongly colored Turnbull's blue:

$$3 \ Fe^{2+} + 2 \ K_3[Fe(CN)_6] \rightarrow Fe_3[Fe(CN)_6] + 6 \ K^+ \tag{3.99}$$

The hydroxyl ions formed during dioxygen reduction can be detected with phenolphthalein as a pH indicator.

A ferroxyl indicator solution contains both reagents in an aqueous solution of NaCl. The salt addition increases ionic conductance, and in addition chloride ions accelerate corrosion and enable a more rapid detection of the proceeding reactions.

Execution
Chemicals and instruments
Ferroxyl indicator solution (3 g NaCl, 0.1 g phenolphthalein and 0.1 g $K_3[Fe(CN)_6]$ in 100 ml water)

Steel sheet

Iron nail

Abrasive paper

Ethanol

Magnifying glass

Procedure
A drop of the indicator solution is put on the cleaned and degreased steel plate. Local color changes are observed with the magnifying glass. Drops of the solution are put onto places of the cleaned and degreased iron nail which have been subject to cold deformation during manufacturing (head, underside of head, tip). Again local color changes are observed with the magnifying glass.

Evaluation
The intense blue color visible already after a few minutes indicates iron dissolution, and the more slowly developing pink coloration indicates formation of hydroxyl ions. On the steel plate metal dissolution dominates in the center of the drop-coated surface; at the nail, places subject to particularly high stress during forming show high rates of iron dissolution.

Literature
H. Kaesche: Corrosion of metals, Springer, Berlin 2003.

Experiment 3.31: Passivation and Activation of an Iron Surface [30]

Task
The cementation reaction of copper ions on an iron surface, the corrosion of iron in concentrated nitric acid, and the passivation of iron in this acid are studied. A mechanical perturbation of this passivation is observed.

Fundamentals
A piece of iron immersed in an aqueous solution of copper ions will be covered by copper deposits, and at the same time iron is dissolved according to

$$Cu^{2+} + Fe \rightarrow Fe^{2+} + Cu \tag{3.100}$$

---

[30] In this experiment aspects of the electrochemical series (Expt. 2.1), of metal deposition (Expt. 7.12 etc.), and of corrosion are united. For a better understanding the pertinent descriptions may be considered.

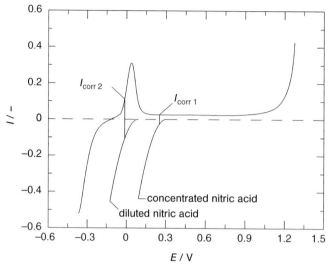

Fig. 3.73 Simplified current-potential curves of an iron electrode and the respective hydrogen electrodes in dilute and concentrated nitric acid.

The thin copper layer is easily dissolved in concentrated nitric acid. Iron is stable in this solution. In contrast iron is not stable in dilute nitric acid. The passivity in the former solution is the result of the establishment of a corrosion potential at the iron electrode in the passive region of iron. The very small corrosion current of the iron electrode is associated with a small current caused by the reduction of protons from the nitric acid (I_{corr1}). Iron is stable (i.e. passivated) in this solution. This is illustrated in Fig. 3.73.

In dilute nitric acid the potential of the hydrogen electrode is shifted according to the lower proton concentration to more negative values, the established corrosion potential is in the region of active iron dissolution (I_{corr2}).

When the piece of iron is lifted in its passivated state from the concentrated nitric acid and immersed in the solution of copper ions no metal deposition is observed. A short mechanical perturbation (a mechanical shock) changes the surface state, and instantaneously metal deposition proceeds again.

Execution
Chemicals and instruments
Aqueous solution of $CuSO_4$ 1 **M**
Concentrated nitric acid
Large iron nail
2 small test tubes
Abrasive paper
Ethanol

Setup
The test tubes are filled with the respective solutions.

Procedure
The cleaned and degreased nail is immersed for a few seconds in the copper sulfate solution. As expected the red color of deposited copper indicates the cementation reaction. The nail is withdrawn and immersed in the concentrated nitric acid. The copper plating is dissolved, and violent foaming and development of nitrogen oxides (use fume hood!) are observed. When the nail is withdrawn and immersed in the copper sulfate solution again no copper cementation occurs. The nail is withdrawn from this solution. A mechanical shock (e.g., a careful hit of the tweezers holding the nail (not the nail covered with solution itself) onto a hard surface) changes the passive surface state of the nail sufficiently to cause sudden copper deposition from the solution adhering to the nail. If the nail is mechanically "disturbed" during handling, this release of passivation ("activation") may occur prematurely.

Experiment 3.32: Cyclic Voltammetry with Corroding Electrodes

Task
The electrochemical response of simple steel (tool steel) and stainless steel is investigated with cyclic voltammetry.

Fundamentals
Cyclic voltammetry employed as a versatile method of electrochemical investigation has already been used in numerous previous experiments. Recording a CV of a nickel electrode in contact with an electrolyte solution showing corrosion and passivation (see Expt. 3.11) was among them. In the following experiment the behavior of tool steel and stainless steel is studied comparatively in a corroding technical electrolyte solution[31].

31) Concentrated solutions of ammonium nitrate are employed as fertilizer in agriculture.

32) When large pieces (thick wire, rod, strips) of steel are used the surface not to be in contact with the electrolyte solution should be covered with adhesive or PTFE tape.

Execution

Chemicals and instruments

Aqueous solution of NH_4NO_3 76wt%

Acetate buffer pH=4.7

Tool steel electrode [32)]

Stainless steel electrode

Platinum wire counter electrode

Reference electrode (mercurous sulfate)

H-cell

Potentiostat, function generator and X-Y-recorder or computer with ADDA-converter and software

Purge gas nitrogen

Setup

The setup for cyclic voltammetry described in Expt. 3.11 is used.

Procedure

The metal strips are cleaned with abrasive paper and degreased. A surface area of about 2 cm² should be exposed to the electrolyte solution. The ammonium nitrate electrolyte solution (with 2 ml of buffer added to 100 ml of electrolyte solution) is filled into the H-cell. The strip electrode is mounted as the working electrode; counter and reference electrodes are inserted respectively. After saturating the electrolyte solution with nitrogen, CVs are recorded from the spontaneously established corrosion potentials to an upper potential limit of $E_{MSE}=1.5$ V.

Evaluation

A typical CV obtained with tool steel is displayed in Fig. 3.74, the corresponding plot with stainless steel is displayed in Fig. 3.75.

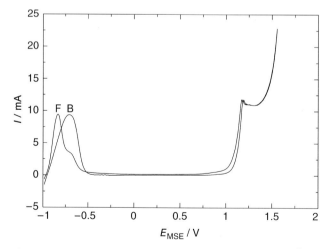

Fig. 3.74 Cyclic voltammogram of a tool steel electrode ($A=2$ cm²) in an aqueous solution of ammonium nitrate, $dE/dt=10$ mV·s⁻¹.

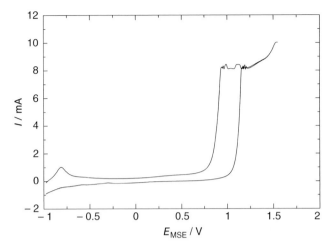

Fig. 3.75 Cyclic voltammogram of a stainless steel electrode ($A=2$ cm^2) in an aqueous solution of ammonium nitrate, $dE/dt=10$ mV·s^{-1}.

In the positive-going scan in Fig. 3.74 active metal dissolution can be observed with a current peak at F. The current drops to very low values in the passive region; vigorous dioxygen evolution follows in the transpassive region. In the negative-going scan the electrode turns back to the active state yielding the current peak at B. The stainless steel electrode shows distinctly different behavior; only in the positive-going scan is some metal dissolution observed; in subsequent scans (not displayed here) the current drops further.

Literature
H. Kaesche: Corrosion of metals, Springer, Berlin 2003.
H. Gerischer, Angew. Chem. **70** (1958) 285.

Experiment 3.33: Oscillating Reactions

Task
Changes in electrode potential and current during an oscillating reaction are recorded and interpreted based on a model of the proceeding reactions.

Fundamentals
Chemical systems showing spatial or temporal structures remote from the thermodynamic equilibrium are called dissipative systems. Temporal structures, stationary spatial structures, or spatial-temporal oscillations may be observed. In addition to a distinctly nonequilibrium situation ($\Delta G \neq 0$), further conditions

must be met: the system must have at least two unstable states, a coupling between various reaction steps, and at least one nonlinear reaction step (autocatalysis, autoinhibition) must be present. Oscillating reactions can be observed easily at the phase boundary metal/electrolyte solution, since the major condition (distinctly nonequilibrium situation) can be met easily because of the small extent of the interface as compared to the volume of the cell vessel.

In this experiment, oscillations of the electrode potential of a copper electrode in a strongly acidic solution of HCl with a current applied are studied. The anodic dissolution of copper can be observed under favorable conditions when visible "schlieren" with a large content of copper ions gliding down from the copper electrode. In the presence of chloride ions the following comproportionation reaction according to

$$Cu^{2+} + Cu + 2\ Cl^- \rightarrow 2\ CuCl \tag{3.101}$$

can proceed causing a white coloration of the copper surface. The high rate of formation of copper ions results in a partial displacement of copper ions at the phase boundary and an associated rise of the local pH value. At a sufficiently high pH value, passivation of the copper electrode with formation of a copper oxide surface layer is possible:

$$2\ Cu + H_2O \rightarrow 2\ Cu_2O + 2\ H^+ + 2\ e^- \tag{3.102}$$

Simultaneously the electrode potential rises steeply. This rise in potential is the precondition for the start of another competitive reaction yielding further copper oxide:

$$Cu_2O + H_2O \rightarrow CuO + 2\ H^+ + 2\ e^- \tag{3.103}$$

This oxide is black, and accordingly the copper surface turns darker; in addition the electrode potential drops to about half the fromer peak value. Copper(II) oxide has poor passivating properties as compared to copper(I) oxide: it is easily dissolved in acid

$$CuO + 2\ H^+ \rightarrow Cu^{2+} + H_2O \tag{3.104}$$

The layer of copper(I) chloride formed during this potential drop will be dissolved yielding a soluble complex:

$$CuCl + Cl^- \rightarrow [CuCl_2]^- + H_2O \tag{3.105}$$

The copper electrode is now unprotected again and turns into the active state showing copper dissolution.

Execution
Chemicals and instruments
Aqueous solution of HCl 5 **M**
Copper electrode
Platinum wire counter electrode
Reference electrode (any type)
Beaker
Adjustable current source
Ampere meter
Adjustable resistor (100 ΩU, 2 A)
Y-t-recoder or transient recorder or other data recording/logging system
Abrasive paper

Setup
The platinum electrode and the cleaned copper electrode are inserted in the beaker filled with the hydrochloric acid solution. The platinum electrode is connected to the negative terminal of the power supply, and the copper electrode is connected via the adjustable resistor and the ampere meter to the positive terminal. The positive terminal of the potential recording device is connected to the copper electrode, and its negative terminal is connected to the reference electrode inserted into the beaker. Depending on the input resistance of the recorder the reference electrode potential may shift. Because only relative potential changes will be recorded without attention to absolute values this is of no importance. If correct absolute values are to be measured, a high input impedance device, if necessary with a input impedance converter, will be necessary. The arrangement is shown in Fig. 3.76.

Procedure
The adjustable resistor is set to $R=30\,\Omega$, and the power supply to about $U=4.45$ V. The most suitable value depends, in addition to other factors on the electrode geometry and distance. Accordingly, currents of a few tens up to a few hundreds of milliamperes are observed. If visible and measurable potential oscillations do not start the value R must be changed until the oscillations start. A

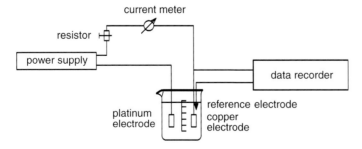

Fig. 3.76 Experimental setup for the study of oscillating reactions.

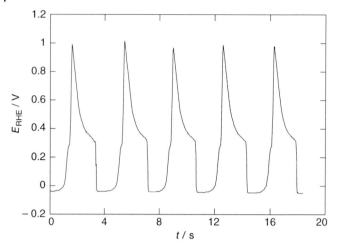

Fig. 3.77 Typical potential-time plot of a copper electrode in an aqueous solution of 5 **M** HCl.

typical potential-time plot recorded with a simple X-Y-recorder is displayed in Fig. 3.77.

Literature

M. Oetken, Praxis der Naturwissenschaften Chemie **47** (1998) 12.
U. Franck, Angew. Chem. **90** (1978) 1.

4
Analytical Electrochemistry

Electrochemical methods and instruments are indispensable in practically all areas of analytical chemistry. In addition to the high sensitivity inherent because of the validity of Faraday's law and the extremely low limits of detection, in many cases easy application, simple connection to other instruments and systems and to data acquisition and processing systems, and the mostly compact design suitable also for mobile applications are attractive features.

The electrochemical phenomena and processes upon which these analytical methods are based are very versatile, and accordingly a systematic organization is difficult. Basically, a distinction can be made between methods employing electrochemical processes for a quantitative determination of analyte and methods where electrochemical phenomena and processes are only employed to detect the point of equivalence in, e.g., a titration. Still, a clear separation and assignment of every method is difficult. The grouping and organization shown below (Fig. 4.1) has turned out to be helpful:

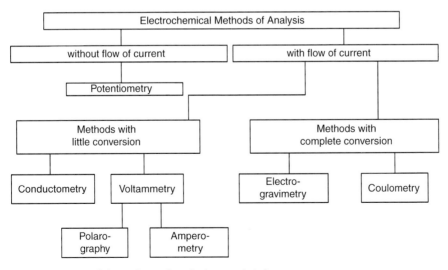

Fig. 4.1 Overview of electrochemical methods in analytical chemistry.

Experimental Electrochemistry. A Laboratory Textbook. Rudolf Holze
Copyright © 2009 WILEY-VCH Verlag GmbH & Co. KGaA, Weinheim
ISBN: 978-3-527-31098-2

According to this overview, potentiometric methods (without a flow of current) are treated first. Then methods where changes in the bulk of the solution sample are important (conductometrically indicated titration) are presented. Finally, methods where charge transfer at the electrode surface is of central importance are handled. The initially discussed fundamental distinction between "completely electrochemical" and "partially electrochemical" methods is not pursued further because it is untenable.

Further details of the treated methods are provided in introductory monographs and handbooks:

Untersuchungsmethoden in der Chemie, H. Naumer and W. Heller Eds., Georg Thieme Verlag, Stuttgart, 2nd ed, 1990.
G. Henze und R. Neeb, Elektrochemische Analytik, Springer Verlag, Berlin 1986.
Analytikum, VEB Deutscher Verlag für Grundstoffindustrie, Leipzig 8th ed., 1990.
D. A. Skoog and J. J Leary, Principles of Instrumental Analysis, Saunders Coll. Publ., Fort Worth, 4th ed, 1992.
M. Geißler, Polarographische Analyse, VCH, Weinheim 1981.
Electroanalytical Methods, F. Scholz Ed., Springer-Verlag, Berlin 2002.

Experiment 4.1: Ion-sensitive Electrode

Task
A silver ion-sensitive electrode is prepared from a graphite electrode and silver sulfide powder. Its applicability is tested by recording a calibration curve with silver nitrate solution, by determining an unknown silver ion concentration, and by using the electrode as an indicator in a precipitation titration.

Fundamentals
In addition to the possibility of preparing ion-sensitive electrodes (ISE) of various types with respective metallic substrates, there is a simple alternative of preparing an ISE without metallic components. Their advantage is obvious: no corrosion of metallic components, no undesirable side reactions by, e.g., dissolution of a substrate. The ISE studied here is prepared by pressing a graphite rod embedded in resin (e.g., PTFE) into a very poorly soluble salt (e.g., Ag_2S, AgCl, CuS) containing the ion to be measured. This way enough salts stick to the moderately hard and electrochemically inert graphite to provide the desired ion sensitivity. In this experiment silver sulfide is used. The electrode obtained can be applied in the determination of sulfide and silver ions. Its usefulness is illustrated with the calibration curve shown in Fig. 4.2. The slope of 51 mV per decade of silver ion concentration deviates from the theoretically expected value, but taking into account the very simple setup it is certainly acceptable.

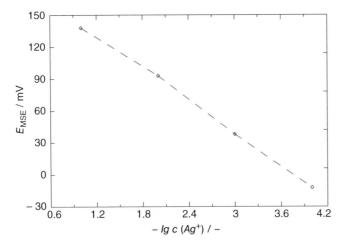

Fig. 4.2 Calibration curve of a graphite electrode impregnated with Ag$_2$S.

Execution
Chemicals and instruments
Aqueous AgNO$_3$ stock solutions (0.1 M, 0.01 M, 1 mM, 0.1 mM)
Aqueous AgNO$_3$ solution of unknown concentration
Aqueous KCl solution (0.1 M) for titration
Ag$_2$S powder
Abrasive powder (Al$_2$O$_3$) 3 μm particle size [1]
Graphite electrode
High input impedance voltmeter
Mercurous sulfate reference electrode

Setup
The reference electrode is connected to the "Low/Common" input of the voltmeter, the ISE to the "High" input. The ISE is prepared by polishing the axial surface of the graphite rod with Al$_2$O$_3$, rinsing with water, and pressing the rod into Ag$_2$S powder.

Procedure
ISE and reference electrode are immersed in the calibration stock solution of known silver ion concentration. The voltage observed after it becomes stable is recorded. The solution of unknown silver ion concentration is reated in the same way. A titration is performed by taking 10 ml of the solution of unknown composition diluted with 40 ml of pure water. The cell voltage is recorded after every addition of 0.5 ml of the KCl-solution.

---

1) Instead fine abrasive paper can be used also.

Evaluation

A calibration curve is obtained by plotting the recorded cell voltages (E_{MSE}/mV vs. $\lg c_{AG^+}$) according to the Nernst equation. By interpolation the unknown concentration is determined. The titration is evaluated by plotting the cell voltage as a function of the added volume of KCl solution.

Literature

H. Galster, Chemie für Labor und Betrieb **36** (1985) 118.
H. Wenk and K. Höner, Chemie in unserer Zeit **23** (1989) 207.

Question

Why is a mercurous reference electrode used instead of a saturated calomel electrode?

Experiment 4.2: Potentiometrically Indicated Titrations

Tasks

- According to a DIN procedure the chloride content of paper is determined.
- With a silver electrode, chloride and iodide are titrated simultaneously.
- With a glass electrode, phosphoric acid is titrated by stepwise neutralization of the tribasic acid.

Fundamentals

Direct potentiometry – i.e. the determination of an analyte concentration directly from the measured potential of the indicator electrode using the Nernst equation – is of limited precision only (about $\pm 1 \ldots 5\%$ because of the logarithmic

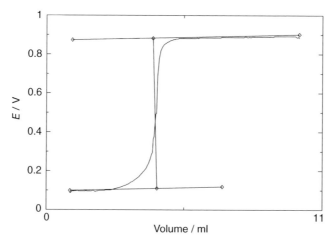

Fig. 4.3 Graphical evaluation of a titration curve yielding its turning point and thus the equivalence point.

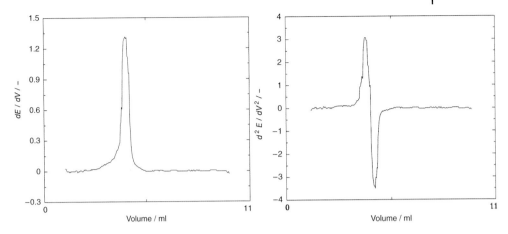

Fig. 4.4 Graphical evaluation of a titration curve yielding its equivalenc point using the first (left) or the second (right) derivative.

relationship) and thus only of limited used. In demanding applications the use of an ion-selective electrode as an indicator electrode in titrations combines the high precision of a titration with the ease of recognition of the equivalence point by electrochemical methods.

In the following three-part experiment this is examined in the complete process of chloride determination in paper, in the simultaneous titration of chloride and iodide ions, and in the acid-base titration of phosphoric acid as a typical example of a polybasic acid.

In titrations of sample solutions with low analyte concentrations, detection of the turning point in the titration curve can be difficult. Figure 4.3 shows a typical example of the graphical evaluation. When the titration curve is available in digital or digitized form, the first or the second derivative can be employed to find the equivalence point (Fig. 4.4).

Execution
Chemicals and instruments
Aqueous $AgNO_3$ solution 0.0025 **M**
Barium nitrate
Acetone
Semiconcentrated nitric acid
Chloride- and iodide-containing solutions of unknown concentration
Phosphoric acid of unknown concentration
Aqueous NaOH solution 0.01 **M**
Silver single-rod electrode[2)]
pH-glass electrode
High input impedance millivoltmeter
Buret

Support stand
Nutsch filter, porosity 4
Water-jet pump
Filtering flask
Round-bottom flask with reflux condenser
Beakers
Magnetic stirrer plate and bar
(Mercurous sulfate reference electrode)

Setup

The buret used for the titrations filled with the titration solution is mounted at the support stand above the magnetic stirrer plate at a suitable height, enabling dropwise addition of the solution into the beaker without the buret tip touching the liquid surface at moderate speed of rotation and without spilling drops.

Procedure

Chloride determination in paper
Two samples of five grams of dry paper, each cut into small pieces, are heated with 100 ml water each for one hour. The extract is obtained by filtering the cooked mass with the nutsch filter. After cooling down of the extract, 50 ml of solution are transferred into another beaker, and a few drops of nitric acid and 50 ml of acetone are added. The single-rod electrode is mounted at a suitable level low enough to provide contact between sample liquid and external reference electrode junction. Titration is done in small steps (0.1 to 0.2 ml per step). Further details are provided in the procedure DIN 53125. The back titration described therein is particularly useful alternative method.

Simultaneous determination of chloride and iodide ions
5 ml of the sample solution of unknown composition are diluted to 50 ml with distilled water. For improved detectability of the equivalence point about one gram of barium nitrate is added. The beaker is placed on the magnetic stirrer plate, and the stirrer bar is added carefully. The single-rod electrode is mounted at a suitable level low enough to provide contact between sample liquid and external reference electrode junction. Titration is done in small steps (0.2 to 0.5 ml per step).

---

2) If none is available a simple silver ion-sensitive electrode
(a silver wire) can be used as an indicator electrode. The reference electrode must be halide-free; a mercurous reference electrode is particularly suitable.

Titration of phosphoric acid
5 ml of the sample solution of unknown composition are diluted to 50 ml with distilled water. For improved detectability of the equivalenc point about one gram of barium nitrate is added. The beaker is placed on the magnetic stirrer plate, and the stirrer bar is added carefully. The single-rod glass pH electrode is mounted at a suitable level low enough to provide contact between sample liquid and external reference electrode junction but sufficiently far away from the stirrer bar to avoid damage of the glass membrane. Titration is done in small steps (0.2 to 0.5 ml per step). Usually only the first and the second neutralization step can be observed.

Evaluation
1. The measured electrode potentials of the silver indicator electroe are plotted as a function of the volume of added titration solution in part a), the equivalence point is derived from the plot. Figure 4.5 shows the titration curve and its first and second derivative.

The figure convincingly demonstrates that an equivalence point derived from the maximum of the first derivative depends on the method applied to obtain this derivative. The graphically determined one is closer to the correct value of the equivalence point.

Alternatively, the equivalence point can be obtained from the second derivative. The volume of titration solution added up to the point where the second

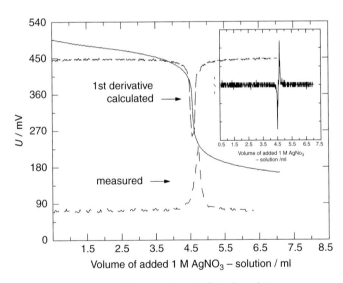

Fig. 4.5 Titration curve of a potentiometrically indicated titration of chloride ions with a solution of silver nitrate; the automatically determined first derivative (motor-driven buret) and the graphically determined first derivative as well as the second derivative of the graphically determined curve (inset) are shown in addition.

derivative passes zero is taken as the volume needed for complete titration. Based on the stoichiometry of the titration reaction, the unknown chloride content is calculated, and the result is related to the weight of the investigated paper. The result is given in grams of chloride per kilogram of paper.

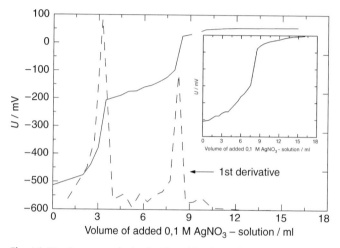

Fig. 4.6 Titration curve obtained with a chloride- and iodide-containing solution during a potentiometrically indicated titration starting with a few milliliters of solutions containing KCl and KI. Inset: Same experiment without added BaNO₃.

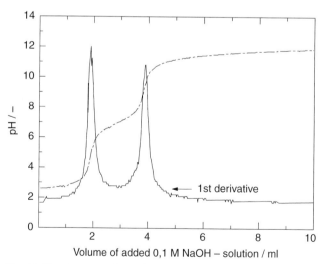

Fig. 4.7 Titration curve obtained with phosphoric acid during a potentiometrically indicated titration.

2. During the simultaneous titration of chloride- and iodide- containing solutions, a titration curve with two steps as shown below in Fig. 4.6 is obtained.

The influence of the added $BaNO_3$ is obvious; without its addition only a single step indicating complete titration of both halides is obtained.

3. During the titration of phosphoric acid a curve with several steps showing at least the first and second step of neutralization is obtained (Fig. 4.7).

Question
Why are only the first and second step of neutralization of phosphoric acid observed?

Literature
DIN 53125.

Experiment 4.3: Bipotentiometrically Indicated Titration

Task
The iodine content of a sample is determined by titration with thiosulfate. The equivalence point is indicated bipotentiometrically.

Fundamentals
Iodine present in aqueous iodide-containing solution as I_3^- can be titrated with thiosulfate. The following reaction proceeds:

$$I_2 + 2\ S_2O_3^{2-} \rightarrow 2\ I^- + S_4O_6^{2-} \tag{4.1}$$

The equivalent point can be detected bipotentiometrically. Across two small indicator electrodes (most simply two small platinum wire electrodes with small surface) a low current (typically a few microamps) is passed. As long as there is iodine in addition to iodide in the solution at both electrodes, reactions according to the redox equilibrium proceed:

$$I_2 + 2\ e^- \leftrightarrows 2\ I^- \tag{4.2}$$

According to the small current the overpotentials at both electrodes are small, the deviation from the rest potentials is low, and the voltage measured between the electrodes is also small. Beyond the equivalence point the iodine concentration drops, and the redox equilbrium is disturbed. Instead of one of the redox reactions of Eq. 4.2 another reaction prevails:

$$2\ S_2O_3^{2-} \rightarrow S_4O_6^{2-} + 2\ e^- \tag{4.3}$$

This reaction proceeds in one direction only. A counter reaction needed to balance this current (e.g. hydrogen evolution) occurs only at significantly more negative electrode potentials. Thus the voltage between the indicator electrodes rises dramatically at the equivalent point.

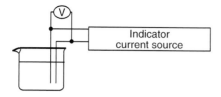

Fig. 4.8 Setup for bipotentiometrically indicated titration of an iodine-containing solution with thiosulfate.

Execution
Chemicals and instruments
Aqueous solution of $K_2S_2O_3$ 0.1 **N**
Aqueous solution of I_2 0.1 **N** with equimolar addition of KI
Current source
Voltmeter
2 Platinum wire electrodes
Buret
Magnetic stirrer plate
Magnetic stirrer bar

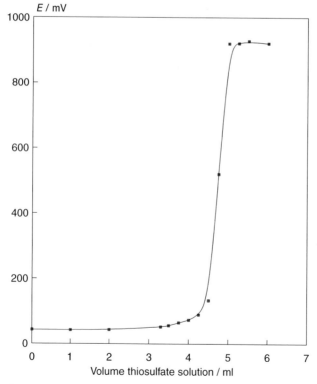

Fig. 4.9 Titration curve of 5 ml of a solution of I_2 with a solution of $K_2S_2O_3$ 0.1 **N**.

Setup

The indicator electrodes are immersed in the sample solution and connected to the current source set to a current of a few microamps as shown in Fig. 4.8.

Procedure

The titration solution is added dropwise to the vigorously stirred sample solution. The voltage between the indicator electrodes is recorded.

Evaluation

A typical titration curve obtained with a sample solution of 0.5 ml of 0.1 **N** iodine solution is shown in Fig. 4.9. From the turning point of the titration curve the equivalence point and hence the composition of the sample solution can be obtained.

Literature

Z. Galus: Fundamentals of Electrochemical Analysis, Ellis Horwood, Chichester 1994.

Experiment 4.4: Conductometrically Indicated Titration

Task

The sulfate content of a sample is determined by a precipitation titration with Pb^{2+} ions. In a redox titration the arsenate(III)[3] content is determined by reaction with iodine.

Fundamentals

Direct concentration determination of ionic species in solution based on the relation between concentration and conductance is hardly suitable for precise quantitative determination[4] whereas conductance measurements are helpful in detecting an equivalence point. The following two examples going beyond the popular acid-base titrations are studied.

In a precipitation titration sulfate ions react with Pb(II) ions forming the poorly soluble $PbSO_4$ according to

$$Pb^{2+} + 2\,NO_3^- + 2\,Na^+ + SO_4^{2-} \rightarrow 2\,Na^+ + 2\,NO_3^- + PbSO_4 \tag{4.4}$$

Up to the equivalence point sulfate ions are replaced by nitrate ions with approximately equally good conducting properties. Beyond this point the excess of lead and nitrate ions yields an increasing conductance (see Fig. 4.10 left).

---

3) Also called arsenite.
4) In some simple devices offered for e.g. pH-value determination of soil this relationship is employed, results are very approximate only at best.

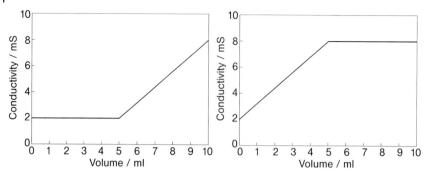

Fig. 4.10 Titration curve of a precipitation titration (left) and a redox titration (right).

In a redox titration the consumption of the substance to be determined or of the titrand can also be monitored conductometrically provided that highly conducting species are formed or consumed. In the example studied here, arsenate(III)(AsO_3^{3-}) is titrated with an alcoholic iodine solution. Up to the equivalence point the conductance of the solution increases because of the formation of protons and iodide ions; beyond this point no further ions are formed. The resulting typical plot is shown in Fig. 4.10 right. The reaction proceeds according to

$$AsO_3^{3-} + I_2 \rightarrow AsO_4^{3-} + 2\,H^+ + 2\,I^- \tag{4.5}$$

Execution
Chemicals and instruments
Aqueous solution of $PbNO_3$ 0.5 **M**
Aqueous solution of unknown concentration of Na_2SO_4
Alcoholic solution of iodine 0.1 **M**
Solution of arsenate(III) of unknown concentration
Conductometer
Conductomeric cell
Buret
Beaker
Magnetic stirrer plate
Magnetic stirrer bar

Setup
The conductometric cell connected to the conductometer is mounted at the support stand low enough to be immersed in the sample solution in the beaker on the stirrer plate without being hit by the stirrer bar.

Procedure

10 ml aliquots of the solutions of unknown concentration are poured into the beaker, and pure water is added up to about 50 ml. Conductance as a function of added titrant is recorded. Because only relative changes of conductance are needed, the cell constant does not need to be known. Conductivities or resistivities as observed at the instrument are sufficient. In the case of the precipitation of $PbSO_4$, after each addition of titrant a delay may be necessary until a stable value appears. Close to the equivalence point relative changes of conductance will be larger, and thus volumes of added titrand should be smaller.

Evaluation

Conductance (or resistivity) values are plotted for both experiments as a function of added volume of titrant. Taking into account the stoichiometry of the titration reaction the unknown concentrations are calculated.

Questions

- Why is careful temperature control needed in measurements of absolute conductivities?
- Why is this control not necessary in the present experiment?

Literature

Z. Galus: Fundamentals of Electrochemical Analysis, Ellis Horwood, Chichester 1994.

Experiment 4.5: Electrogravimetry

Task

The copper content of a sample solution is determined by complete electrolytic deposition as copper metal.

Fundamentals

An electrical current passed through a solution containing metal cations causes deposition of the corresponding metal ions at the electrode connected to the minus pole (negative electrode, cathode). Only non-noble metals (e.g. sodium, potassium) placed in the electrochemical series negative to the hydrogen electrode are not deposited; instead, hydrogen is evolved.

As a cathode a platinum net electrode is used. Most metals will form well-adhering deposits on this. Provided all metal ions from the sample solution are deposited and their mass is determined by weighing, the method is called electrogravimetry. Platinum, as a very noble electrode material, permits the deposited metal, after weighing, to be selectively dissolved with nitric acid. As an anode also platinum (a wire coil) is used.

During electrolysis of a solution of $CuSO_4$, in addition to cathodic copper deposition, formation of oxygen is observed:

Cathode (reduction): $Cu^{2+} + 2\ e^- \rightarrow Cu$ $\qquad$ (4.6)

Anode (oxidation): $H_2O \rightarrow 2\ H^+ + 1/2\ O_2 + 2\ e^-$ $\qquad$ (4.7)

Total: $Cu^{2+} + H_2O \rightarrow Cu + 2\ H^+ + 1/2\ O_2$ $\qquad$ (4.8)

Copper ions are replaced by protons, and an amount of sulfuric acid equivalent to the amount of copper sulfate is formed.

When the platinum electrode is coated with copper it is formally converted into a copper electrode, i.e. its potential depends on the copper ion concentration according to the Nernst equation. As the copper ion concentration decreases during deposition, the electrode potential is changed, and accordingly the cell voltage needed to continue deposition has to be increased. (Which increase of cell voltage would it be necessary to increase in order to decrease the copper ion concentration to zero?). An electrogravimetric determination is considered to be finished when the cathode potential is shifted in the negative direction by about $\Delta E \approx -200$ mV; this is equivalent to a negligible residual concentration. In order to maintain high ionic conductance during electrolysis the sample solution is acidified with sulfuric acid. Mass transport to the electrodes is enhanced by vigorous stirring; heating to about 50 °C further enhances the rate of deposition.

Execution
Chemicals and instruments
Aqueous solution of unknown concentration of $CuSO_4$
Sulfuric acid 20 wt%
Semiconcentrated nitric acid
Platinum wire net cathode, platinum wire anode
Current source
2 Multimeters
Beaker 250 ml
Magnetic stirrer plate
Magnetic stirrer bar

Setup
The setup is schematically shown in Fig. 4.11.

Procedure
The sample solution is heated in the beaker to 50 °C. The carefully cleaned and degreased (with alcohol) weighed platinum net electrode and the platinum wire anode are fixed at a support stand at a height where both electrodes are immersed. They are connected via the multimeters operated as voltmeter and ammeter to the adjustable current source.

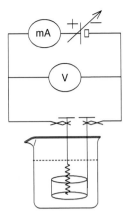

Fig. 4.11 Setup for electrogravimetry.

A current of $I=100$ mA is applied. Slow decoloration of the solution and a red coating on the cathode indicate the progress of electrolysis. Towards the end of copper deposition the applied voltage rises significantly, indicating that instead of copper deposition undesirable hydrogen evolution is setting in. After complete deposition (discoloration) the electrodes are removed from solution with the current source still attached (why?). They are carefully rinsed with water. After turning off power the net electrode is rinsed with ethanol and dried. Its weight is carefully determined.

Finally the copper deposit is etched away with semiconcentrated nitric acid. After complete removal, the electrode is rinsed first with water, then with alcohol, and dried.

Evaluation
From the weight difference the amount of copper and finally the mass of copper sulfate in the initial sample are determined.

Questions
- What change of electrolysis voltage do you expect when switching off current?
- Why does the electrolysis voltage increase slowly during the experiment at constant stirring?
- Why is the solution heated during electrolysis?

Experiment 4.6: Coulometric Titration

Task

The content of As^{3+} ions in a solution is determined by reaction with electro-generated bromine with biamperometric indication of the equivalence point.

Fundamentals

Transformation of reactants at an electrode can be calculated according to Faraday's Law provided the electrode reaction is known and well defined. During electrolysis at constant current the consumed charge and thus transformed amount of matter is equivalent to the product of current and time. Because even small currents can be determined with high precision, high sensitivity is possible. The resulting methods are collectively known as galvanostatic coulometry.

In these methods, wether the electrochemical reaction itself or the product are employed to determine quantitatively the content of a species in a solution. The second case is called coulometric titration. The electric current is used as titrant. An important condition for a successful application is that the electrochemical generation of the titrant proceeds at perfect yield (no parasitic reactions); in addition, the end of the titration (when the species to be determined are completely consumed) must be detected easily.

End points (or equivalen points) are preferably detected using electrochemical (potentiometric or amperometric) methods. During potentiometrically indicated titrations the change of potential during the titration yields the equivalence point: during amperimetrically indicated titrations, changes in the current flowing through indicator electrodes yield this information. All coulometric methods are absolute ones, i.e. no weighing, preparation of standard solutions etc., are necessary. As an example, the suggested coulometric titration is discussed in detail below. The aim is the coulometric determination of As^{3+} with electrolytically generated bromine according to

$$As^{3+} + Br^2 \rightarrow As^{5+} + 2\ Br^- \tag{4.9}$$

The sample solution is acidified, and potassium bromide is added. Two platinum ("generator") electrodes are used for the generation of bromine

$$\text{Cathode: } 2\ H^+ + 2\ e^- \rightarrow H_2 \tag{4.10}$$

$$\text{Anode: } 2\ Br^- \rightarrow Br_2 + 2\ e^- \tag{4.11}$$

The generator current circuit is operated at constant current. The anodically generated bromine is immediately consumed by the reaction with As^{3+} ions in solution. To keep bromine formed at the anode from reaching the cathode and being converted into bromide again, the cathode is separated from the stirred solution by a porous glass frit. At the equivalene point free bromine stays in the solution and is no longer cosumed. It is detected by two platinum ("indicator") electrodes. A small voltage ($U = 500$ mV) applied at these electrodes is insufficient to cause electrolysis of the solution. Only when free bromine is available is the reversible redox system

$$2\ Br^- \rightleftarrows Br_2 + 2\ e^- \tag{4.12}$$

with negligibly small overvoltage established, and a significantly large current flows. Thus the equivalent point is detected by the sudden rise of the current in the indicator circuit because the electrodes are now depolarized. Because two indicator electrodes are employed the method is also called biamperometry.

The mass of As^{3+} ions in the sample is calculated according to Faraday's Law

$$m(As^{3+}) = M(As)((I \cdot t)/(2 \cdot F)) \tag{4.13}$$

Execution
Chemicals and instruments
0.1 **N** Aqueous stock solution of As^{3+}
0.1 **N** Aqueous solution of H_2SO_4
0.2 **N** Aqueous solution of KBr
Beaker with generator and indicator electrodes
2 Ammeters (mA and μA)
Stop watch
Volumetric flask 10 ml
Pipet 20 ml
Volumetric cylinder 25 ml
Magnetic stirrer plate
Magnetic stirrer bar

Setup
The setup is schematically shown in Fig. 4.12:

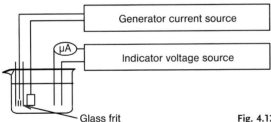

Fig. 4.12 Setup for coulometric titration.

Procedure
A sample solution with a concentration of 0.005 **N** As^{3+} is prepared from the stock solution. The electrodes and the glass tube closed with a frit at the bottom are mounted in the still empty beaker at a level keeping them well immersed (when the beaker is filled) and allowing free movement of the stirrer bar. 25 ml of 0.1 **N** H_2SO_4 and 25 ml of 0.2 **N** KBr are poured into the beaker, and distilled water is

added until the beaker is about half full. The glass tube is filled immediately up to the same liquid level with 0.1 N H_2SO_4. The current adjustment at the generator current supply is set to zero, and the voltage at the indicator voltage source is set to $U = 500$ mV. The generator current is set to exactly $I = 5$ mA. 1 ml of the As^{3+} sample solution is added, and the stop watch is started. When the current in the indicator circuit rises the watch is stopped; the experiment is finished. All used solutions are disposed of carefully because of the arsenic content.

Evaluation

The amount of As^{3+} is calculated and compared with that one expected from the known solution composition. Sensitivity and precision of the methods may be estimated.

Questions
- Define the terminology "galvanostatic coulometry", "coulometric titration" and "biamperometric titration".
- State the electrode reactions in the generator and the indicator circuit.
- Name further examples of coulometric determinations.

Experiment 4.7: Amperometry

Tasks
In amperometrically indicated titrations
- the content of lead ions is determined by precipitation with dichromate ions
- the content of styrene is determined by bromatometric titration

Fundamentals

In all titrations, detection of the equivalence point is a central task. Electrochemical methods are fairly independent of experimental conditions and the expertise and cognitive capabilities of the experimentator. In addition electrical signals are generated which can be processed, transformed, and transmitted easily.

When during a titration an electrochemically active reducible or oxidizable compound is formed whose concentration shows a significant change at the equivalence point, the current needed for this transformation can be measured and used as a measure of the progress of the titration. Small electrodes and correspondingly small currents result in only very minor consumption of species and thus negligible errors of measurement. This can be illustrated in the following example during the titration of lead ions with sulfate ions:

$$Pb^{2+} + SO_4^{2-} \rightarrow PbSO_4 \tag{4.14}$$

Lead ions, which can be reduced easily at a sufficiently negative electrode potential, are consumed during the titration, and thus the reduction current decreases. Beyond the equivalent point only a small residual current flows. Sulfate ions are electrochemically inactive at the applied electrode potential. Titration of sulfate ions with lead ions results in a reverse dependence:

$$SO_4^{2-} + Pb^{2+} \rightarrow PBSO_4 \tag{4.15}$$

The current is initially very small, butbeyond the equivalence point it grows rapidly because of the excess of lead ions. In another variation, both the species to be determined and the titrant are electrochemically active. Lead ions can be precipitated with dichromate ions:

$$2\ Pb^{2+} + Cr_2O_7^{2-} + H_2O \rightarrow 2\ PbCrO_4 + 2\ H^+ \tag{4.16}$$

Both ions can be reduced cathodically:

$$Pb^{2+} + 2^- \rightarrow Pb \tag{4.17}$$

$$Cr_2O_7^{2-} + 14\ H^+ + 6\ e^- \rightarrow 2\ Cr^{3+} + 7\ H_2O \tag{4.18}$$

The cathodic current due to the reduction of lead ions will thus decrease up to the equivalence point; beyond ths the current will rise again because of the re-duction of excess dichromate ions. The shape of the obtained curve depends on the type of titrand and the electrode potential applied to the indicator electrode. For three conceivable combinations examples are shown and discussed below.

Up to the equivalence point lead ions are reduced, but beyond this point di-chromate ions are reduced (Fig. 4.13).

In a second case (see Fig. 4.14), up to the equivalence point the current is al-most zero because the electrode potential is not sufficiently negative to support reduction of lead ions. Beyond the equivalence point reduction of dichromate ions proceeds. A similar shape of curve is observed when the species to be de-termined is electrochemically inactive (i.e. in the titration sulfate ions with lead ions) and the potential is sufficient to support transformation of the titrant (in this example $E_{SCE} = -1\ V$).

Current / mA

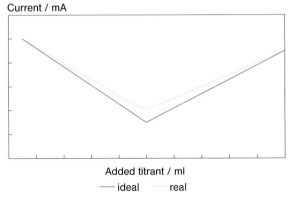

Added titrant / ml

—— ideal ········ real

Fig. 4.13 Titration curve of a precipitation titration of lead ions with dichromate ions at $E_{SCE} = -1\ V$.

Current / mA

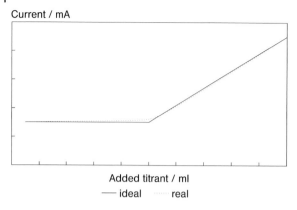

Added titrant / ml
—— ideal ‥‥‥ real

Fig. 4.14 Titration curve of a precipitation titration of lead ions with dichromate ions at $E_{SCE}=0$ V.

Current / mA

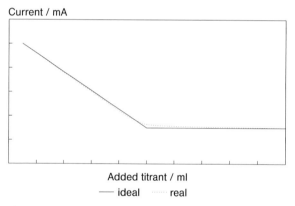

Added titrant / ml
—— ideal ‥‥‥ real

Fig. 4.15 Titration curve of a precipitation titration of dichromate ions with lead ions at $E_{SCE}=0$ V.

In a third case, up to the equivalence point dichromate ions are reduced, ut beyond the equivalence point the potential of the indicator electrode is not sufficiently negative to support lead ion reduction (Fig. 4.15).

A similar curve may be observed with an electrochemically inactive titrant, e.g., in the precipitation of lead ions with sulfate ions. In this case the indicator electrode potential must be sufficiently negative to support reduction of lead ions ($E_{SCE}=-1.0$ V).

The method requires an arrangement where the indicator electrode can be maintained at a well-defined electrode potential. Because the current depends on the type and concentration of the species to be transformed as well as he reproducible mass transport, constant conditions must be established during the experiment. Use of a supporting electrolyte (to avoid the influence of migration), an electrode potential in the diffusion limited region and constant mass transport established by, e.g., stirring are suggested.

Hanging mercury drop electrodes or rotating platinum disk or pin electrodes have been used successfully for reducible species; for oxidations only, platinum can be used because mercury is oxidized quite easily. A stationary platinum electrode may be suitable also in most cases instead of a rotating one. To set the desired indicator electrode potential a reference electrode (e.g., a saturated calomel electrode) and a three-electrode arrangement with a potentiostat are basically required. Because only small currents are flowing when an indicator electrode with very small surface area (as recommended) is used, an unpolariz-able reference electrode with low internal impedance (most saturated calomel electrodes fit this requirements) can be used as both reference and counter elec-trode in a two-electrode arrangement. Since dioxygen can also be reduced in most cases in addition to the species to be measured, purging of the sample so-lution with nitrogen or other inert gas is required before every current measure-ment and after addition of titrant.

It is even possible to employ amperometric detection without an external volt-age source by using the voltage appearing between indicator and reference elec-trode. When, e.g., a platinum indicator electrode is used connected with a calo-mel electrode via a sensitive ammeter, no current will appear when at the spon-taneously established electrode potential neither the titration products nor the substance to be determined can be electrochemically transformed. When the titrant can be transformed under these conditions its excess appearance beyond the equivalent point will cause a flow of current.

Amperometric indication is also useful in bromatometric titration of styrene. Into the sample solution with added KBr and methanol acidified with HCl a platinum wire electrode is immersed as indicator electrode. Its potential with re-spect to a saturated calomel electrode is about $E_{SCE}=0.25$ V. This potential does not support any electrode reaction at the platinum wire (when indicator and ref-erence electrode are connected via an ammeter). Upon addition of bromate ions this does not change as long as bromine formed according to

$$5 \text{ Br}^- + \text{BrO}_3^- + 6 \text{ H}^+ \rightarrow 3 \text{ Br}_2 + 3 \text{ H}_2\text{O} \tag{4.19}$$

is consumed by reaction with styrene

$$\tag{4.20}$$

Only when free bromine appears in solution will a current based on the revers-ible reaction

$$\text{Br}_2 + 2 \text{ e}^- \rightleftharpoons 2 \text{ Br}^- \tag{4.21}$$

flow.

Execution

Chemicals and instruments

0.01 **M** Aqueous solution of $Pb(NO_3)_2$

Aqueous solution with unknown concentration of $Pb(NO_3)_2$

0.001 **M** Aqueous solution of $KBrO_3$

0.0005 **M** Aqueous solution of $K_2Cr_2O_7$

0.2 **M** Aqueous solution of NH_4NO_3

Methanol

KBr

Concentrated hydrochloric acid

Aqueous solution of styrene [5]

Ice

DC voltage source

Ammeter (μA)

Dropping mercury electrode

Saturated calomel electrode

Platinum tip electrode

Nitrogen gas

Beaker 250 ml

Volumetric cylinder 100 ml

Volumetric cylinder 10 ml

Buret

Magnetic stirrer plate

Magnetic stirrer bar

Setup

1. Lead determination

The mercury electrode and the calomel electrode are connected to the DC voltage source via the ammeter. Stirring of the solution is effected by the gas purge.

2. Titration of styrene

The platinum tip electrode and the saturated calomel electrode are connected via the ammeter. The sample solution is magnetically stirred.

Procedure

1. Lead determination

Currents should be read immediately before each drop falls, and somewhat longer drop times (about 2 s) are suggested. Before every measurement dissolved dioxygen is purged with nitrogen.

First a current-potential curve is registered in order to identify a suitable electrode potential for the subsequent titration. To 50 ml of the NH_4NO_3 solution

---

[5] This solution can be prepared by vigorous shaking of water with styrene (styrene is a severe health risk – use a fume hood) and subsequent decanting or separation in a separation funnel. Formation of an opaque solution indicates formation of an emulsion; after some setting phase separation will occur.

1 ml of the $Pb(NO_3)_2$ solution is added. After purging the dioxygen the current-potential curve is registered in 50 mV steps in the range 0 ... −1200 mV. Subsequent recording of the curve for the dichromate ion reduction requires complete precipitation of lead ions from the sample solution. This can be accomplished by adding about 15 ml of the dichromate solution. The current-potential curve is registered for dichromate reduction as before for lead ions. From the obtained curves suitable electrode potentials for the following titration are selected.

For the titration 1 ml of the lead ion solution of unknown concentration is added to 50 ml of the NH_4NO_3 solution in the beaker. Electrodes are fixed, and the selected electrode potential is applied. After purging with nitrogen, dichromate solution is added in 0.5-ml steps. After every addition the solution is purged, and the peak current at the end of the drop life is recorded. The current will be reproducibly constant only when the solution has settled after purging. The titration is repeated with titrant addition in 1-ml steps. Does the addition of larger amounts affect the precision?

2. Titration of styrene
75 ml of methanol and 5 ml of concentrated HCl are filled into the beaker, which is cooled in an ice bath to 5 ... 10 °C. The styrene-containing sample (here: 25 ml of the saturated solution) is added. If the temperature has risen above 10 °C, cooling with ice is again required. After addition of 1 g of KBr the solution of $KBrO_3$ is used as titrant. A current range of 20 µA at the ammeter is suggested.

Evaluation
The equivalence points are determined from the obtained plots. The unknown lead ion concentration should be reported in mol/l. Stoichiometry of the reactions and molarity of the employed solutions must be carefully observed.

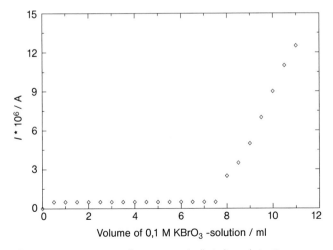

Fig. 4.16 Titration curve of amperometrically indicated titration of styrene with bromate/bromide.

From the value obtained in the titration of styrene its solubility in water is calculated. Because styrene is rather volatile without special precautions in most cases too low values are obtained. In the reported example a styrene content of 0.0093 wt% is found. The literature value is 0.023 wt% (D. H. James and W. M. Castor in: Ullmann's, 5th ed., 1994, Vol. A25, p. 330).

Questions

- Which factors must be observed when selecting the voltage for the indicator circuit? On which influences does the current depend?
- Explain the shape of the observed titration curves.
- Estimate the error caused by the cathodic lead deposition. Assume 5 min time is needed to reach the equivalene point with a solution containing 3 mg lead ions and an average current in the indicator circuit of 2.5 µA.

Experiment 4.8: Polarography (Fundamentals)

Tasks

- Determine the half-wave potentials of copper, cadmium, and zinc.
- Determine for cadmium ions the dependence of the diffusion-limited current (height of the polarographic wave) on the concentration for use in quantitative determination.
- Determine the composition of a sample solution of unknown composition (qualitatively: copper, cadmium, zinc; quantitatively: cadmium).

Fundamentals

Polarography is a voltammetric method. Voltammetry (a short form of *voltamperometry*) is generally the measurement of the current flowing through the test electrode in the electrochemical cell as a function of the electrode potential of the voltage applied to the cell. A specialty of polarography is the use of a mercury electrode. As a counter electrode, various non-polarizable electrodes may be employed; the reaction at the counter electrode is almost always of no interest. The advantages of a mercury electrode and in particular a dropping one are:

- Frequent renewal of the electrode surface
- The high hydrogen overpotential (i.e. kinetic inhibition of hydrogen evolution by electrolyte solution decomposition) extends the usable negative range of electrode potentials down to $E_{Ag/AgCl} = -1.8$ V.

At the mercury working electrode reduction or oxidation processes can only occur when an electrode potential sufficiently negative or positive to drive conversion of solution species is applied. This electrode reaction results in depletion of the electrolyte solution in the region close to the mercury surface, and a concentration gradient increasing in a direction towards the solution is established.

This gradient drives mass transport by diffusion. Mass transport by artificial convection is excluded in the stagnant solution. Natural convection, conceivable because of local changes in solution density caused by the depletion of the solution, is negligible, because the amount of species converted is small and thus density changes are small also. Migration caused by the electric field between the electrodes is suppressed by addition of a large concentration of electrochemically inert supporting electrolyte (KCl, NH₄Cl). At a sufficiently large electrode potential, all species arriving at the electrode surface are immediately converted; their surface concentration is zero. The resulting current is called diffusion-limited current I_{diff}.

According to Fick's first law the rate of diffusion is proportional to the concentration gradient $\partial c/\partial x$ at the interface. In the diffusion-limited case the surface concentration is $c_s = 0 \ mol \cdot l^{-1}$; now the concentration gradient can be described simply by the thickness of the Nernstian diffusion layer δ_N and the bulk concentration c_0 of the species (Fig. 4.17):

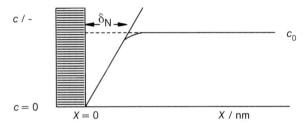

Fig. 4.17 Concentration profile at the electrochemical interface.

This provides a possibility to determine the concentration of the species converted at a given electrode potential from I_{diff}. Further means of mass transport have been excluded by suitable precautions. Dioxygen generally present in solutions exposed to air must be purged with inert gas because it will be reduced at sufficiently negative electrode potentials, thus distorting the recorded current-potential curves. In a dioxygen-free electrolyte solution with supporting electrolyte only (e.g., 1 M NH₄Cl), a negative current will appear at very negative electrode potentials because of the large hydrogen overpotential of mercury and because the reduction of the ammonium cation requires a very negative electrode potential (see trace 1 in Fig. 4.18).

Upon addition of, e.g., Zn^{2+}, Cd^{2+} and Cu^{2+}, ions trace 2 will be obtained. A simple electronic smoothing of the curve with a low pass filter yields the curve shown in the inset, which is easier to evaluate because the trace appears more clearly. The recorded current-voltage curves oscillate with the dropping frequency of the dropping mercury electrode. The average value of I_{diff} during a drop lifetime is given by the Ilkovič equation:

$$\bar{I}_{diff} = 607 \cdot n \cdot \sqrt{D} \cdot m^{2/3} \cdot c_0 \cdot \tau^{1/6} \tag{4.22}$$

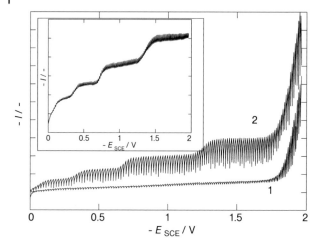

Fig. 4.18 Simple polarograms [6] of an aqueous solution of
1 M $NH_4Cl + 0.5$ M NH_4OH (1) and with 3 mg Cu^{2+}, Cd^{2+}
and Zn^{2+} per ml added (2); inset with slight electronic
smoothing.

with:
D: diffusion coefficient in $cm^2 \cdot s^{-1}$
M: flow rate of mercury at the dropping mercury electrode in $mg \cdot s^{-1}$
τ: dropping time in s

The electrode potential observed at half height of the step $(I = I_{diff}/2)$ is a typical
property of a species. With reversible electrode reactions where the Nernst equation is valid, the half wave-potential equals the standard potential of the electrode reaction. Because a time-dependent DC voltage is employed the method is
called DC-polarography.

Execution
Chemicals and instruments
2 M Aqueous solution of NH_4Cl
1 M Aqueous solution of NH_4OH
Standard solutions with $1\ mg \cdot ml^{-1}$ of Cu^{2+}, Cd^{2+}, and Zn^{2+}
Sample solution of unknown copper, cadmium, and zinc content
Dropping mercury electrode
Saturated calomel electrode
Polarographic cell [7]
Triangular voltage sweep generator (or equivalent voltage source)
Digital voltmeter

6) According to polarographic convention axis are scaled contrary
to the usual manner.
7) For very simple experiments a beaker might be sufficient.

XY-recorder
Buret
Nitrogen purge gas

Setup

The basic setup is shown schematically in Fig. 4.19. Dropping mercury electrode and polarographic cell are placed in a drip tray to collect mercury in case
of failure of a component or leakage.

Procedure

To 50 ml of the supporting electrolyte solution prepared by mixing 200 ml of
the 2 M solution of NH_4Cl and 100 ml of the 1 M solution of NH_4OH, cadmium ion-containing solution resulting in 1.25, 2.5, 3.75 and 5 mg Cd^{2+} in
50 ml solution are added. The voltage generator is set to a range 0 ... –2 V (with
respect to the reference electrode conveniently used instead of the mercury pool
electrode at the cell bottom), and the scan rate is set to $dE/dt = 10$ mV·s^{-1}. The
resistor used as a shunt between voltage source and mercury capillary to record
the current is set to 500 Ω.

The supporting electrolyte solution is filled into the cell and purged with nitrogen for 5 min. The mercury reservoir is fixed approx. 35 cm above the cell
(actual height might vary) resulting in a drop rate of 1 s^{-1} when the valve is
opened.

Recording of polarograms

The following polarograms are recorded:
1. 50 ml of supporting electrolyte solution
2. the same solution with 3 ml of copper ion solution added

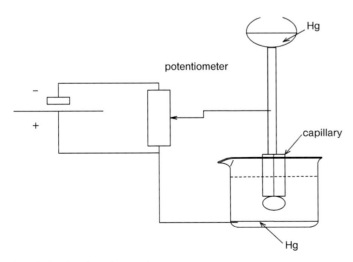

Fig. 4.19 Setup for polarography.

3. previous solution with 3 ml of cadmium ion solution added
4. previous solution with 3 ml of zinc ion solution added
5. 1.25 mg Cd^{2+} in 50 ml solution
6. 2.5 mg Cd^{2+} in 50 ml solution
7. 3.75 mg Cd^{2+} in 50 ml solution
8. 5 mg Cd^{2+} in 50 ml solution
9. sample solution of unknown copper, cadmium and zinc content

Evaluation

From polarogram 1 the decomposition voltage of the supporting electrolyte solution is determined. Polarograms 2 ... 4 yield the half-wave potentials of the respective metals. From polarograms 5 ... 8 a calibration curve for cadmium with the height of the step vs. the concentration is plotted. From polarogram 9 the presence of the various metals and the concentration of cadmium ions (use the calibration curve) is derived.

Questions

- List the advantages of a dropping mercury electrode in voltammetry.
- What is a polarogram?
- Describe the function of the supporting electrolyte.
- What roles do migration, diffusion and convection lay in the polarographic cell?
- Describe and explain a typical polarogram.
- How is the half-wave potential related to the standard potential?
- How are polarograms evaluated qualitatively and quantitatively?
- Why does the drop rate depend on the electrode potential?
- How can cadmium-containing solutions from these experiments be disposed of properly?

Experiment 4.9: Polarography (Advanced Methods)

Task

Heavy metal ions at low concentrations are determined quantitatively.

Fundamentals

Under the term "polarography" various analytical voltammetric techniques are collected which all employ a mercury working electrode. In all cases the current flowing through this electrode is recorded as a function of the applied electrode potential. This current is caused by an electrode reaction (reduction or oxidation) proceeding at the working electrode. In the case of a simple two-electrode system (as in the previous experiment) the potential of counter electrode must be well defined and constant because this electrode serves also as the reference electrode. Typical curves obtained by DC-polarography are shown above (see

Fig. 4.18). A more detailed description of other polarographic methods is provided in the literature (EC:494).

Execution
Chemicals and instruments
Aqueous solution 0.01 **M** $Pb(NO_3)_2$ in 0.1 **M** KCl
Aqueous solutions with 2.5 $mg \cdot ml^{-1}$ of Pb^{2+}, Cu^{2+}, Cd^{2+}, and Zn^{2+}
Aqueous acetate buffer solution pH = 4.5
Polarograph for DC, differential pulse and tast polarography
Dropping mercury electrode
XY-recorder
Pipet 10 ml, 20 ml
Buret 2 ml
Micro syringe 100 μl
8 volumetric flasks 100 ml

Setup
The polarograph is connected with the recorder and the polarographic cell in a three-electrode arrangement.

Procedure
1. I_{diff} of the reduction of lead ions is determined with the aqueous solution 0.01 **M** $Pb(NO_3)_2$ in 0.1 **M** KCl at various flow rates of mercury (given as drop time τ).

2. The concentration of Cu^{2+}, Cd^{2+}, Pb^{2+}, and Zn^{2+} in an aqueous acetate buffer solution pH = 4.5; methods: differential pulse polarography (solutions a .... d) and sampling polarography (solution d)
   a) 10 ml buffer solution
   b) +100 μl of metal solutions each
   c) +200 μl of metal solutions each
   d) +300 μl of metal solutions each
   e) solution from d) with sampling polarography
   f) solution of unknown concentration

Evaluation
1. Determine the half-wave potential and the height of the polarographic wave as a function of the drop time taking into account the Ilkovič equation.

2. Determine peak potentials and peak heights. Plot a calibration curve for the various heavy metal ions. Determine the composition of the solution of unknown composition (qualitatively and quantitatively).

Literature
R. C. Kapoor and B. S. Aggarwal: Principles of polarography, Wiley, New York 1991.

Experiment 4.10: Anodic Stripping Voltammetry[8]

Tasks

Using a mercury film electrode, the composition of a solution containing small metal ion concentrations is determined. For comparison by standard addition, references in terms of identity and concentration of these ions are established.

Fundamentals

To increase sensitivity and in particular to lower levels of detection various methods have been proposed (EC:494). Besides procedures to minimize unwanted contributions from capacitive charging currents, preaccumulation by cathodic deposition and subsequent anodic dissolution of the amalgams thus formed provide dramatic improvements of the level of detection (down to 0.001 ppb). As an alternative to preaccumulation at the hanging mercury drop (obviously a dropping mercury electrode cannot be used), from a vigorously stirred solution mercury film electrodes on graphite supports have been employed with great success. In this approach the mercury film and thus the amalgam are formed simultaneously during the preaccumulation of the studied metal ions at a graphite or glassy carbon electrode. This method works advantageously without handling larger amounts of the environmentally problematic mercury and also avoids difficulties handling a hanging mercury drop. Finally the formed film is completely dissolved during the anodic scan during the actual concentration determination, and cleaning the electrode is reduced to a simple rinse.

After preaccumulation by applying a sufficiently negative electrode potential just above hydrogen evolution analysis proceeds by anodic oxidation. In the most simple case a slow anodic potential scan is enough. Further improvement can be achieved by applying one of the various advanced polarographies (e.g., pulse polarography etc.). According to their position in the electrochemical series, the deposited metals are dissolved from the amalgam sequentially. In contrast to a classic polarogram (step-like diffusion-limited current waves), anodic current peaks are observed. This is because of the changed transport conditions. The height of the peak can be given for a planar electrode by the Randles-Ševčik equation in good approximation:

$$I_p = 2.69 \cdot 10^5 \cdot A \cdot n^{3/2} \cdot \sqrt{D} \cdot \sqrt{v} \cdot c_0 \tag{4.23}$$

For a hanging mercury drop the equation derived by Nicholson and Shain is valid

$$I_p = 602 \cdot n^{3/2} \cdot A \cdot \sqrt{D} \cdot [0.4463 + 0.160 \cdot (1/r)/(D/(n \cdot v))]c_0 \cdot \sqrt{v} \tag{4.24}$$

---

8) This method is also called anodic stripping polarography.

These non-trivial relationships make direct correlations between concentration of a species and peak height difficult, so instead calibration curves or preferably standard additions are employed.

Execution
Chemicals and instruments
Aqueous solution 0.1 **M** KCl
Aqueous solutions of 5 g·l^{-1} Hg(NO$_3$)$_2$ or HgCl$_2$
Electrochemical cell (e.g., H-cell)
Glassy carbon electrode
Saturated calomel reference electrode
Platinum wire counter electrode
Potentiostat and function generator
XY-recorder
Magnetic stirrer plate
Magnetic stirrer bar

Setup
Potentiostat, cell, and recorder are wired like in standard voltammetry. The following example is based on 20 ml solution in the cell.

Procedure
A small measured amount of the sample solution containing the cations of unknown identity and concentration is added to the electrolyte solution (aqueous

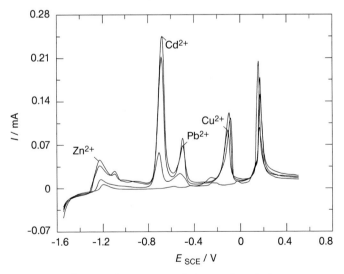

Fig. 4.20 Single scan voltammograms during anodic stripping voltammetry; bottom trace: supporting electrolyte solution; next trace: addition of sample solution; further traces after standard addition as described in text.

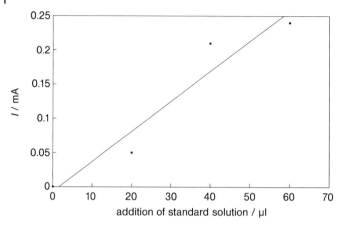

Fig. 4.21 Calibration curve for cadmium ion addition.

solution 0.1 **M** KCl) in the cell. 0.1 ml of the mercury-containg solution are added. With intense stirring the electrode is kept 10 min at $E_{SCE}=1.5$ V to deposit a mercury film containing the accumulated reduced cations. Starting at this potential, a slow scan up to mercury dissolution ($E_{SCE}=0.25$ V) is recorded at a scan rate of $dE/dt=20$ mV·s^{-1}. After addition of standard solutions (here: 20 µl each of solutions containing 2.5 mg·ml^{-1} of copper, cadmium, lead, and zinc ions), the process is repeated twice.

Evaluation
A typical set of voltammograms is shown in Fig. 4.20.

A plot of peak current vs. concentration (as, e.g., for cadmium ions in Fig. 4.21) permits determination of the unknown concentration; the identity of the ions is already obvious in this case.

Literature
R. S. Nicholson and I. Shain, Anal. Chem. **36** (1964) 706.
A. J. Bard and L. R. Faulkner: Electrochemical Methods, Wiley, New York 2001, p. 459.

Experiment 4.11: Abrasive Stripping Voltammetry

Tasks

The composition of metal alloys is determined qualitatively with ASV[9]. For comparison the pure metals are studied.

Fundamentals

Scratching a hard graphite electrode (in the most simple case a pencil lead will suffice) over a metal surface is enough to transfer traces of the metal to the graphite surface. With this graphite electrode used as a working electrode in a three-electrode arrangement with a suitable electrolyte solution (which does not promote passivation of the assumed metals) in an anodic scan these metal traces can be dissolved oxidatively. The electrode potential at which oxidation occurs depends on the position of the metal in the electrochemical series. Less noble metals are dissolved first, more noble ones at more positive electrode potentials. Very noble metals are not dissolved at all, and cannot be determined by this method.

The anodic potential scan can be performed most easily with the setup known from cyclic voltammetry (compare Expt. 3.11). The formation of current peaks can be explained easily, as with stripping voltammetry, taking into account hindered mass transfer and limited supply of metal atoms from the small amount of metal sample transferred by the scratching procedure. As a working electrode a pencil lead coated with a heat-shrinkable plastic[10] and equipped with an electrical connector is used.

Figure 4.22 shows a set of typical CVs obtained with pure lead and antimony as reference materials and with alloys of various compositions prepared by melting different amounts of the metals.

Alloy 1 contains lead and antimony at equal mass percents, in alloy 2 are 66 wt% lead and 34 wt% antimony. The shift of the peak potentials where metal dissolution from the alloys is observed as compared to the pure metals is obvious; it must be caused by slight differences between the properties of the metal in its pure form and its properties as an alloy.

Execution

Chemicals and instruments

Aqueous solution 0.1 M NH_4NO_3
Samples of copper, lead, tin, silver
Alloy sample of unknown composition
Pencil lead working electrode
Saturated calomel reference electrode

9) Instead of the acronym ASV already employed for anodic stripping voltammetry the acronym AbrSV (*Abrasive* stripping voltammetry) has been proposed.

10) A heat-shrinkable tube is made from extruded and expanded polyolefin which when heated shrinks to a fraction (1/3 or 1/4) of its former diameter and clings closely to the wrapped object providing very good insulation.

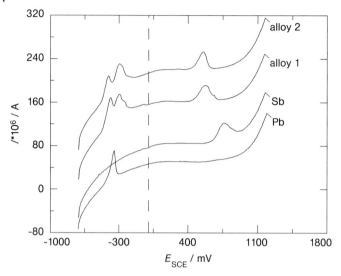

Fig. 4.22 Linear scan voltammograms of a lead pencil electrode in an aqueous electrolyte solution of 0.5 M NH$_4$NO$_3$, dE/dt=50 mV·s^{-1}.

Platinum wire counter electrode
H-cell
Potentiostat and function generator
XY-recorder
Nitrogen purge gas

Procedure

The H-cell is filled with aqueous solution 0.1 M NH$_4$NO$_3$, and the electrodes are fitted and connected to the potentiostat. To record a CV of the blank working electrode the solution is purged with inert gas to remove dioxygen. A CV is recorded in the potential range − 0.7 V < E_{SCE} < 1.5 V, i.e. between the limits at which hydrogen and dioxygen evolution should start.

The graphite electrode is scratched with as much pressure as possible (without breaking) it across the metal surface to be investigated. The function generator is set to its negative potential limit, and the graphite electrode is inserted and wired. A first positive-going scan is started with dE/dt=25 mV·s^{-1}. In a second complete scan cycle complete metal dissolution can be verified. Purging with inert gas is not necessary because dioxygen is electrochemically inactive in the studied range of electrode potentials, and in addition the gas stream might tear away metal particles adhering to the graphite electrode. The procedure is repeated with metal samples of known composition and identity.

The alloy in Fig. 4.23 is a dental amalgam containing silver, copper, and other metals.

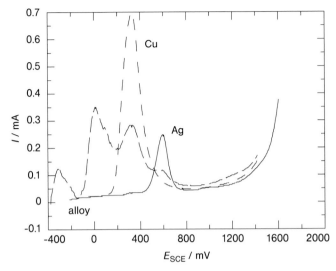

Fig. 4.23 Linear scan voltammograms of a lead pencil electrode in an aqueous electrolyte solution of 0.5 **M** NH$_4$NO$_3$, with traces of copper, silver, and a metal alloy, dE/dt=25 mV·s^{-1}.

Evaluation

By comparison of peak potentials obtained with the sample and with known metals or with literature data on standard potentials the alloy composition is determined qualitatively.

Literature

F. Scholz, L. Nitschke, and G. Henrion, Naturwissenschaften **76** (1989) 71.
F. Scholz, L. Nitschke, G. Henrion, and F. Damaschun, Naturwissenschaften **76** (1989) 167.
F. Scholz, W.-D. Müller, L. Nitschke, F. Rabi, L. Livanova, C. Fleischfresser, and Ch. Thierfelder, Fresenius J. Anal. Chem. **338** (1990) 37.

Questions

Can this method also be applied to studies of metal compounds?

Why are peak potentials observed during metal dissolution from alloys shifted with respect to the values found with pure metals?

Experiment 4.12: Polarographic Analysis of Anions

Tasks

- Quantitative determination of iodide in table salt.
- Quantitative determination of sulfate in mineral water.

Fundamentals

In trace analysis of anions, polarographic methods can also be employed successfully. Some anions can be reduced directly at a mercury electrode. Examples are NO_3^-, NO_2^-, BrO_3^-, IO_3^-, and IO_4^-. IO_3^- is reduced according to

$$IO_3^- + 6\,H^+ + 6\,e^- \rightarrow I^- + 3\,H_2O \qquad (4.25)$$

The large electrochemical valence (6 electrons per ion) results in high diffusion-limited currents and thus a high sensitivity of this method. This can also be employed in the determination of I^-. Iodide itself yields a polarographic wave, but this is less useful in analytical application. However, the wave caused by iodate reduction is six times larger. Thus iodide is oxidized first into iodate, and this is subsequently determined. This procedure is even possible in the presence of large excesses of chloride. It can be applied in iodide determination in table salt (first task).

The convenient polarographic determination of iodate concentration results in a procedure described by Humphrey for the determination of Cl^-, CN^-, F^-, SO_4^{2-} and SO_3^{2-}. The procedure, also called the "amplification method", is based on the reaction of the anion X^- to be determined with a metal iodate $MeIO_3$ according to

$$MeIO_3 + X^- \rightarrow MX + IO_3^- \qquad (4.26)$$

where MX and $MeIO_3$ are insoluble or undissociated compounds. To the solution with the anion to be determined, a water-ethanol mixture (1:1) with a suitable metal iodate is added:

$Ba(IO_3)_2$ for determination of SO_4^{2-}

$Hg_2(IO_3)_2$ for determination of Cl^-, CN^-, or SO_3^{2-}

The solution is shaken and filtered. An equivalent amount of iodate is released, which can be determined subsequently with polarography. This procedure is employed in the second task.

Execution
Chemicals and instruments
Alkaline aqueous hypobromite solution (50 ml 5 **M** NaOH+50 ml bromine water)
Aqueous saturated solution of Na_2SO_3 (for reduction of excess hypobromite)
Aqueous solution of gelatin 0.25 wt%
Aqueous standard solution of iodide (0.05 g·l^{-1} KI)
Aqueous standard solution of sulfate (1 mg·l^{-1} sulfate)
Ethanol
Barium iodate
Concentrated perchloric acid
Table salt
Iodized table salt
Sodium chloride p.A.

2 Mineral waters with different sulfate contents
Setup for DC polarography
1 Pipet 1 ml
2 Pipets 10 ml
6 Measurement flasks 50 ml
6 Beakers 50 ml
Filtration paper for barium sulfate
Funnel
Nitrogen purge gas

Setup
The setup for DC polarography (see exp. 4.19) is used.

Procedure
Determination of iodide in table salt
Table salt and iodized table salt (with artificially added KI) are tested, and for comparison sodium chloride p.A. is examined. 10 g of the salt are dissolved in boiling water, further water is added up to 50 ml. 10 ml of this sample solution are put into the cell for polarography. 1 ml of water, 1 ml of the hypobromite solution, 0.5 ml of the gelatin solution, and 0.5 ml of the sodium sulfite solution are added. After stirring and purging with nitrogen a polarogram is recorded in the range $-0.5 > E_{SCE} > -1.4$ V. The procedure is repeated with all samples. For comparison, in a further run instead of 1 ml water 1 ml of the standard iodide solution is added.

The iodide content is calculated according to

$$x = 25 \frac{a}{b-a} \text{ (mg KI/kg salt)} \tag{4.27}$$

with step height a obtained with a sample and b obtained with the standard solution.

Determination of sulfate in mineral water
Two mineral waters with significantly different contents of sulfate are studied; as a reference pure water is used. CO_2-containing waters must be brought to a boil first to remove this. After cooling down, 25 ml are transferred into a measurement flask. The flask is almost filled up to the mark, 0.5 g of $BaIO_3$ are added, and further ethanol is added exactly up to the mark. The flask is vigorously shaken for 20 min and the contents are filtered. To the filtrate 0.5 ml concentrated perchloric acid is added. After purging with nitrogen a polarogram is recorded in the range $0.15 > E_{SCE} > -0.5$ V. As a reference a solution with a known sulfate concentration close to the value expected from the information of the provider of the mineral water prepared with the standard solution is studied.

Evaluation

The sensitivity of the method is discussed. Obtained results are compared with data given by the samples manufacturer.

Literature

R.E. Humphrey and S.W. Sharp, Anal. Chem. **48** (1976) 222.

Questions

Explain the purpose and mode of action of the various additives used in both tasks.

In both tasks iodate concentration is determined by cathodic reduction. The half-wave potential differs considerably. Why?

Derive eq. 4.27.

Experiment 4.13: Tensammetry

Task

From measurements of the differential double layer capacity of a hanging mercury drop as a function of electrode potential and of concentration of a dissolved alcohol, the parameters of the Frumkin adsorption isotherm of this system shall be determined.

Fundamentals

In AC polarography a small AC voltage (typical values: $10 \ldots 100$ Hz, $5 \ldots 50$ mV) is superimposed on the DC ramp. From the resulting current the DC component is removed, and only the AC component is amplified, rectified and registered. Accordingly, $I_{\approx}/E$-curves showing current peaks with $E_p = $ peak potential and $I_p = $ peak height instead of current steps as expected from DC polarography are observed.

The peak generation is schematically illustrated in Fig. 4.24.

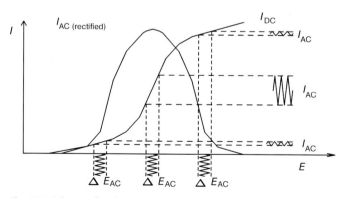

Fig. 4.24 Scheme of peak generation in AC polarography.

The superimposed AC voltage causes a periodic sequence of redox or recharge processes at the given DC potential around this (mean) value. The respective AC currents show a maximum at the steepest point of the $I_\approx/E$-curve of the DC polarogram; they vanish when the DC current stays constant as a function of applied DC potential. Formally speaking, the first derivative of the DC polarogram is obtained. Actually the result is not just a mirror image with respect to the DC polarogram (like, e.g., a characteristic of a vacuum tube or a transistor). In the case of more or less irreversible electrode reactions having different half-wave potentials for the reduction and the oxidation step, not all species reduced in the negative part of the AC wave are reoxidized in the positive part and vice versa. In case of a completely irreversible process, nothing will be reoxidized. Thus the height of the peaks of $I_\approx$ depend very much on the kinetic parameters of the electrode reaction. This is a drawback in the quantitative analysis of completely irreversibly reduced species; it can also be an advantage when, e.g., reversibly reoxidizable species are present in small amounts besides an excess of species which cannot be reoxidized. An advantage of AC polarography is the enhanced resolution: the capability to allow distinctions between species with closely spaced half-wave potentials. In this case the DC polarogram will not show a well-defined current step, making determination of the step height difficult or even impossible. In the case of AC polarography, both peaks are measured versus the same baseline. This is a general advantage of all methods showing current peaks instead of current waves, as already observed in Expt. 4.9 with differential pulse polarography. A drawback of AC polarography is the additional charging current caused by the superimposed AC voltage resulting in a lower detection limit around $c=10^{-4}$ M. This capacitive current can be advantageously employed in double-layer studies. Thus, analytical and kinetic studies of surface-active species (tensides) influencing the double layer capacitance and thus $I_\approx$ are possible. These measurements, also called "tensammetry", are the subject of this experiment. They are actually measurements of the capacitive and not of the Faradaic component of $I_\approx$. Accordingly, also substances which are not electrochemically active themselves can be investigated. When such a species is adsorbed at a mercury electrode the double layer capacitance and thus $I_\approx$ is changed as compared to the value observed in the absence of this addition. Further AC current peaks are observed at electrode potentials where adsorption and desorption of these species proceeds (see Fig. 4.25).

In this plot instead of $I_\approx$ the differential double-layer capacitance is plotted, which is proportional to $I_\approx$ according to

$$I_c = \frac{dE}{dt} C_{diff} \tag{4.28}$$

with $dE/dt=v=$const.

The differential double-layer capacitance C_{diff} is related to the integral double-layer capacitance C_{int} with

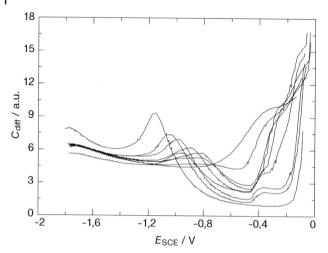

Fig. 4.25 C_{diff} as a function of electrode potential E_{SCE} for various concentrations of *t*-butanol added to the electrolyte solution of 1 **M** KCl: 0.0, 0.04, 0.08, 0.1, 0.15, 0.2, 0.25, 0.3, 0.4 **M** (falling curves at $E_{SCE} = -0.5$ V).

$$C_{diff}(E) = C_{int}(E) + \left(\frac{dC_{int}}{dE}\right) \cdot \left(E - E_{pzc}\right) \qquad (4.29)$$

For simplification, a potential $E^* = E - E_{pcz}$ can be used. E_{pcz} is the electrode potential of zero charge: the point of the electrocapillary maximum (EC:124). Lowering of the C_{diff}/E-curve (see Fig. 4.25) because of adsorption is equivalent to a lower value of C_{int} in Eq. 4.29. C_{int} depends on E and the properties and amount of adsorbed species, whereas dC_{int}/dE in the range $-0.5 < ESC < 0.8$ V is negligible. The peaks corresponding to the adsorption and desorption processes are caused by restructuring of the double layer, e.g., replacement of supporting electrolyte anions and thus the contribution $(dC_{int}/dE) \cdot E^*$. For analytical purposes, these maxima are evaluated. Thus, a method is available enabling the determination of the concentration of electrochemically inactive substances which cannot be measured by other electrochemical methods.

In this experiment the potential range wherein C_{diff} is lowered is evaluated. From this change observed at a selected potential as a function of the concentration of the adsorbable species in solution, the degree of coverage of these species on the surface, θ, can be determined. Assuming C as the value of C_{diff} at a given concentration, C_0 as the value at zero concentration and C_{max} as the smallest value of C where further increase of the concentration does not yield further capacitance reduction (thus $\theta = 1$ can be assumed) θ can be calculated

$$\theta = \frac{C_0 - C}{C_0 - C_{max}} \qquad (4.30)$$

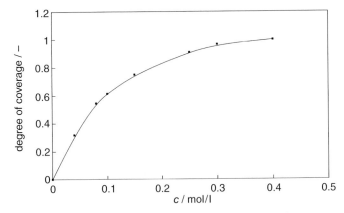

Fig. 4.26 Degree of coverage θ with *t*-butanol as a function of concentration *c*.

A plot of θ vs. the concentration of the adsorbable species yields the adsorption isotherm as shown in a typical example in Fig. 4.26.

When C_{max} is not attained in the studied range of concentrations (i.e. with further increase of concentration the curves show continuously lower minima) a plot of $(C-C_0)/C_0$ vs. concentration with subsequent extrapolation to C_{max} may help.

With graphic evaluation the adsorption coefficient B and the interaction coefficient a of the Frumkin isotherm can be obtained:

$$B \cdot c = \left(\frac{\theta}{1-\theta}\right) e^{-2a\theta} \tag{4.31}$$

A plot of θ vs. $\ln c - \ln(\theta/(1-\theta)$ yields an intercept providing B which can be used to calculate the Gibbs energy of adsorption according to

$$\Delta G_{ad} = (B - \ln 55.5)R \cdot T \tag{4.32}$$

(see Fig. 4.27).

From the illustrated data a value of $\Delta G_{ad} = -14.4 \text{ kJ} \cdot \text{mol}^{-1}$ can be calculated in good agreement with the literature value of $\Delta G_{ad} = -14.0 \text{ kJ} \cdot \text{mol}^{-1}$ (A. de Battisti, B.A. Abd-El-Nabey and S. Trasatti, J. Chem. Soc. Faraday Trans. I **72** (1976) 2076).

Execution

Chemicals and instruments

Alkaline aqueous solution 1 **M** KCl

t-Butanol

Hanging mercury drop electrode with reference and counter electrode in electrochemical cell

Potentiostat

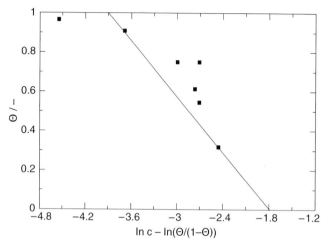

Fig. 4.27 Test of the Frumkin isotherm.

Function generator
Sine wave generator
Frequency selective amplifier
X-Y-Recorder
Heating plate
Microliter syringe
Nitrogen purge gas

Setup
Figure 4.27 shows schematically the setup for tensammetry.

Procedure
The instruments are connected as indicated. First a plot of C vs. E is obtained in the absence of alcohol in order to get C_0. 10 ml of the supporting electrolyte solution are put into the cell, a hanging mercury drop is created, and the solution is purged for 10 min with nitrogen. In the potential range $-1.7 > E_{SCE} > 0.05$ V, a C vs. E plot is recorded at a scan rate $dE/dt = 10$ mV·s^{-1}, a sine wave amplitude of $U_{\approx} = 30$ mV and a frequency of 80 s^{-1}. In steps of 10 µl (up to a total of 100 µl) and of 50 µl (up to a final total of 400 µl) t-butanol is added. After every addition the solution is purged and a plot of C vs. E is recorded. Because t-butanol solidifies at 25.5 °C it is kept warm on the heating plate together with the syringe. In the described setup 1 µl is equivalent to a concentration increase of 10^{-5} M. The additions thus result in concentrations of 0.01 M, 0.02 M, up to 0.1 M, subsequently in steps of 0.05 M to 0.4 M.

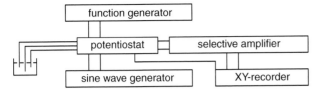

Fig. 4.28 Setup for tensammetry.

Evaluation

From a plot of $(C-C_0)/C_0$ vs. concentration C_{max} is obtained. With this value the adsorption isotherm can be plotted. From the test of the Frumkin isotherm B and ΔG_{ad} can be calculated. The result may be compared with literature values.

Literature

B. B. Damaskin, O. A. Petrii, and V. V. Batrakov: Adsorption of Organic Compounds on Electrodes, Plenum Press, New York 1971.

5
Non-Traditional Electrochemistry

Besides measurements of variables electrode potential, cell voltage, current, charge and their dependence on further parameters like concentration, temperature or pressure, numerous other methods from various branches of natural sciences and technology have been applied in electrochemistry. In particular, spectroscopies and surface-analytical methods have been modified for this purpose, in particular for *in situ* investigation in the presence of an electrolyte solution. This way, numerous open questions impossible to answer with traditional electrochemical methods have been answered. The still growing family of these methods nontraditional in electrochemistry has been named spectroelectrochemistry. In most cases at first glance the need for more expensive instruments appears as a drawback; in most cases spectrometers or comparably large analytical instruments are needed. Only in a few cases, e.g., in measurements of the conductance of polymer films deposited on an electrode, the desired information (here: the dependence of the conductance on electrode potential, electrolyte solution composition, etc.) can be obtained with an extremely simple set-up. The experiments described below thus depend on the availability of the respective instruments. In some cases modifications of these instruments may be needed to perform the suggested experiments, leading to more than the usual scale of preparation for a laboratory experiment for educational purposes. Thus the following texts are suggestions rather than complete descriptions. The spectra are typical examples realistically obtained in a laboratory course; they do not represent necessarily the theoretical optimum.

Experiment 5.1: UV-Vis Spectroscopy

Task

A polyaniline film is deposited on an ITO-electrode[1] and characterized with *in situ* UV-Vis spectroscopy.

1) ITO = Indium-doped tin oxide, transparent and electronically conductive coating on a transparent substrate (glass) which can be used as an electrode. ITO-coated glass is used extensively in liquid crystal displays.

Experimental Electrochemistry. A Laboratory Textbook. Rudolf Holze
Copyright © 2009 WILEY-VCH Verlag GmbH & Co. KGaA, Weinheim
ISBN: 978-3-527-31098-2

Fundamentals

In analytical chemistry spectroscopy of electronic transitions in the ultraviolet and the visible range of light (UV-Vis spectroscopy) is mostly employed in quantitative analysis. In physics this method yields information about electro-optic properties. In electrochemistry both aspects can be employed. There are several experimental approaches differing most prominently in the way the beam of light is guided towards the electrode and the type of electrode used. In external reflection the beam is guided towards the electrode and reflected. Upon reflection, interaction with the electrode surface, modifying layers, adsorbates, etc. may take place resulting in spectral modifications of the reflected light. Even the electrode metal itself might interact; e.g., gold shows its "golden" color because of an intraband transition resulting in optical absorption below 560 nm and reflection of (yellow) light with longer wavelenght, whereas aluminum has no color and is thus closer to the ideal reflector. The reflected beam is guided towards the detector, and comparison with a reference beam (or spectrum) yields an absorption spectrum. As a reference, the electrode at a different electrode potential might be taken, and the obtained differential spectrum thus contains information about electrode potential-dependent changes of electrode surface reflectivity caused by, e.g., adsorbate coverage, structure, and composition of surface films etc., this approach is employed in Expt. 5.3.

In another approach an optically transparent electrode (OTE) and a transmission arrangement in the spectrometer are used. The OTE can be prepared from glass slides coated with indium-doped tin oxide. Such an electrode coated with the material to be studied is inserted into a standard cuvette serving as the electrochemical cell. Counter and reference electrode are placed outside the beam path. This is particularly important when during the experiment optically absorbing products may be formed at the counter electrode. In the reference channel of the spectrometer, a cuvette with the OTE and the electrolyte solution are placed. Thus measurements with a standard two-beam spectrometer yield spectra showing only absorptions caused by matter present on the OTE in the sample cell. Figure 5.1 shows major components of the setup.

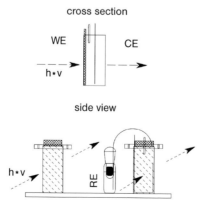

Fig. 5.1 Setup for spectroelectrochemical studies with optically transparent electrodes.

   In the present experiment, further properties of intrinsically conducting polymer, an exciting relatively recent class of materials using polyaniline already studied in Expt. 3.21 as an example, are explored. On an OTE a polyaniline film is deposited. Its absorption spectrum in the presence of perchloric acid is recorded as a function of electrode potential. From the spectra, information about electrooptical properties of the polymer are obtained.

Execution

Chemicals and instruments

Aqueous solution 1 **M** $HClO_4$

Aniline or *o*-toluidine, 0.2 **M** in aqueous solution 1 **M** $HClO_4$

UV-Vis cell with ITO-working electrode, gold wire counter electrode, and reference electrode with salt bridge

UV-Vis spectrometer

Potentiostat

Function generator

Setup

For electropolymerization the same cell also used in the UV-Vis experiments can be employed. Working, counter, and reference electrode are connected to the potentiostat. In subsequent spectroscopic experiments the monomer-containing polymerization solution is substituted by a monomer-free supporting electrolyte solution. In the reference beam of the spectrometer a cuvette with supporting electrolyte solution and an ITO-electrode are placed.

Procedure

In a suitable electrochemical cell or directly in the UV-Vis cuvette, a polymer film is deposited potentiostatically on the ITO-coated glass electrode. The polymerization solution contains 0.2 **M** aniline or *o*-toluidine in the aqueous solution 1 **M** $HClO_4$. The potential is kept at $E_{RHE} = 1$ V until a visible but still transparent film is formed. If no film formation is observed a slightly higher electrode potential should be used. Finally the film is discharged (i.e. reduced; it is formed in its colored, oxidized form) at $E_{RHE} = 0$ V for about 3 min. A change of color is observed. After switching the potentiostat to "stand by" the film is carefully removed form the solution and rinsed with perchloric acid to remove adhering monomers and oligomers. Supporting electrolyte solution is filled into the UV-Vis cuvette, and counter and reference electrode are installed.

   UV-Vis spectra are recorded at various electrode potentials ($E_{RHE} = 0$; 0.1; 0.2 ... 0.9 V). After every potential step a delay period of 2 min is observed (why?); in this time the current should drop to negligible values. Finally a value of $E_{RHE} = 1.5$ V is set, and several spectra of the slowly overoxidizing film are recorded with a few minutes intervals.

   When *o*-toluidine is used instead of aniline the deposition potential is set to $E_{RHE} = 1.05$ V.

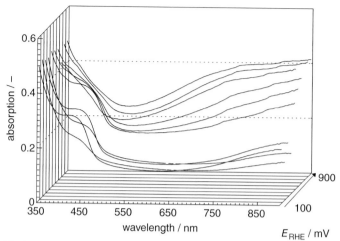

Fig. 5.2 UV-Vis spectra of a polyaniline film in an aqueous solution of 1 M $HClO_4$ as a function of electrode potential. The film was deposited from a polymerization solution of 0.2 **M** + aniline 1 **M** $HClO_4$ at $E_{RHE} = 1$ V.

Evaluation

Spectra are plotted preferably in a 3-dimensional arrangement as illustrated in Fig. 5.2, which hows results obtained with polyaniline in perchloric acid. Wavelength is plotted on the x-axis, absorption on the y-axis; electrode potentials or overoxidation time are plotted from front to back.

The spectrum obtained at low potentials shows a band assigned to electronic transitions into electronic states caused by the formation of radical cations (polarons) during polymer oxidation. The absorption around $\lambda = 600$ nm can be correlated with the electronic conductance of the film, which can also be measured *in situ*. The involved states are spinless dications (bipolarons). These changes provide hints at the identity of the carriers of electric charge in the polymer. With even more positive potentials, absorption extends into the range of longer wavelengths (near infrared NIR), and the maximum moves to shorter wavelength (blue shift) with higher potentials. These changes provide further evidence regarding the involved species. A plot of the absorbance at selected wavelengths as a function of electrode potential is particularly instructive, and should be discussed. Spectra obtained as a function of time should also be discussed with respect to observed changes and new bands and their conceivable origins.

Literature

K. Menke and S. Roth, Chemie in unserer Zeit, **20** (1) (1986) 33.
R. B. Kaner and A.G. MacDiarmid, Scientific American **268(2)** (1988) 106.
P. M. S. Monk, R.J. Mortimer and D.R. Rosseinsky: Electrochromism: Fundamentals and Applications, VCH, Weinheim 1995.

Experiment 5.2: Surface Enhanced Raman Spectroscopy

Task
Surface-enhanced Raman spectra (SERS) of a molecule or ion adsorbed on a coinage metal electrode are recorded and interpreted.

Fundamentals
The identification of adsorbed species on a surface and the study of their inter-action with their environment (solution, electrode surface) is possible with *in situ* vibrational spectroscopies; in particular, Raman and infrared spectroscopies are used (EC:295). As Raman spectroscopy is based on a scattering process with inherently low photon yield, studies of two-dimensional interfaces with even lower numbers of potential scatters would appear fruitless initially. With roughened or rough deposited metal surfaces of the d-metals (coinage metals Cu, Ag, Au) a large surface-enhancement effect (10^6) is observed, which enables routine studies of adsorbates to be carried out. Extension to other metals is pos-sible by depositing these metals as thin, pinhole-free layers on rough d-metal surfaces (gold) or by depositing a thin layer of a d-metal (silver) on the metal to be studied. In infrared spectroscopy such materials limitations do not exist. Because measurements are mostly performed in external reflection, the strong attenuation of the infrared beam by water (the most commonly used electrolyte solvent) and the cell window cause serious problems. This may be overcome by modulation techniques, but the spectra obtained this way may be differential

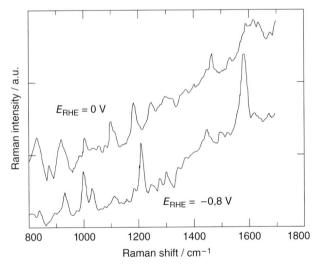

Fig. 5.3 SER spectra of an electrochemically roughened silver electrode in an aqueous solution of 0.1 **M** $HClO_4$ + approx. 1 **mM** pyridine, saturated with nitrogen, electrode potentials as indicated, $\lambda_{Laser}=514.5$ nm, $P_{Laser}=400$ mW, resolution 9 cm^{-1}, spectrometer Dilor RT 20.

ones showing only the differences in electrode reflectivity between the applied electrode potentials, causing uncertainties in interpretation.

A typical example of an SER spectrum is shown in Fig. 5.3.

In the investigated range of electrode potentials at $E_{RHE}=0$ V beyond the band assigned to the symmetric stretch mode ν_4 of the perchlorate anion at 935 cm^{-1} only several weak bands are observed. At the more negative potential $E_{RHE}=$ -0.8 V, bands typical of pyridine are observed at 1003, 1035, 1214, and about 1600 cm^{-1}. Appearance (or disappearance) of bands and changes in intensity and position as a function of electrode potential can be evaluated. This may yield information about the degree of coverage of the silver electrode with pyridine and about the adsorbate geometry. Of particular interest is the electrode potential of zero charge E_{pzc}, at which the coverage with neutral adsorbate molecules tends to be particularly high. The nitrogen atoms play a special role in the adsorptive interaction. For comparison a Raman spectrum of pyridine dissolved in the supporting electrolyte solution is helpful. A spectrum of neat pyridine is less useful, because it does not reveal the effect of interaction with water and electrolyte (acid).

Combination of results obtained with SERS and with *in situ* infrared spectroscopy is particularly powerful, as both methods complement each other. In infrared spectroscopy, surface selection rules enable very definite statements to be made about adsorbate geometry; this is less straightforward in SERS. Instead, SERS is more sensitive.

Execution
Chemicals and instruments
Laser Raman spectrometer
Spectroelectrochemical cell
Potentiostat
Function generator
H-cell
Nitrogen purge gas

Setup
For electrochemical roughening, a procedure described in the literature is employed (R. Holze, Electrochim. Acta, **32** (1987) 1527). After rinsing with pure water, the roughened electrode is transferred into the spectroelectrochemical cell containing the supporting electrolyte solution and the substance to be adsorbed. The cell is then attached to the spectrometer.

Because of the particularly high safety risks involved with the use of lasers, instructions must be carefully observed.

Literature
R. Holze, Electroanalysis, **5** (1993) 497.
R. Holze: Surface and Interface Analysis: An Electrochemists Toolbox, Springer Verlag, Berlin 2009.
Spectroelectrochemistry, (R. J. Gale, ed.), Plenum Press, New York 1988.
Electrochemical Interfaces (H. D. Abruna, ed.), VCH, New York 1991.

Experiment 5.3: Infrared Spectroelectrochemistry

Task
The electrode potential-modulated infrared reflection absorption spectrum of CO_{ad} formed upon chemisorptive interaction between methanol and platinum is recorded as a function of electrode potential.

Fundamentals
As already stated in the description of the preceding experiment, *in situ* vibrational spectroscopies are very powerful methods in detecting and studying organic adsorbates at electrode surfaces. The possibility to identify even unknown adsorbates based on their vibrational spectrum in the so-called fingerprint region well known in classical analytical spectroscopy is frequently employed. The application *in situ*, i.e. in the presence of an electrolyte solution, avoids any conceivable detrimental effect of sample transfer outside the cell possibly causing experimental artifacts. In this example, infrared spectroscopy is applied. In the external reflection mode the beam of light travels through the infrared-transparent cell window and a thin layer of electrolyte solution[2]. Upon reflection at the electrode surface, interaction between the electric field vector of the light (more precisely: the *p*-polarized light) and infrared vibrational modes of adsorbate species may proceed; the reflected beam carrying this additional information is guided again through the solution layer and the window on to the detector. As in classical infrared spectroscopy, two spectra are needed to obtain a result containing only the desired information. In the present experiment this is achieved by recording two so-called single-beam spectra (actually records of reflection vs. wavenumber) at two different electrode potentials designated E_r and E_m. These potentials are selected based on previous electrochemical studies with, e.g., cyclic voltammetry; they should refer to distinctly different states of the electrode in terms of adsorbate properties, coverage etc. Further details are provided in the literature.

Execution
Chemicals and instruments
Aqueous solution 1 **M** $HClO_4$
Methanol
Fourier transform infrared (FTIR) spectrometer
Spectroelectrochemical cell with electrodes
Potentiostat with interface to spectrometer
Nitrogen purge gas

---

2) The well-known problems of thin layer electrochemistry must be kept in mind when assessing the obtained spectra.

Setup

Potential-modulated differential infrared spectra are recorded at electrode potentials as determined from preceding cyclic voltammetry or as proposed by the supervisor. They are most suitably set to a value E_r where only adsorption of methanol proceeds and E_m where this adsorbate shows considerable changes in binding.

Evaluation

A typical example is shown in Fig. 5.4. The obtained spectra are discussed with respect to identity of the adsorbate and influence of the electrode potential on the adsorbate. In the selected mode of display equivalent to the standard transmission mode display, positive (upward) pointing bands indicate higher infrared absorption at E_r, and negative (downward) pointing bands indicate higher infrared absorption at E_m. The differential shape of the band is caused by the fact that its position is shifted slightly between the two electrode potentials, whereas the coverage changes only slightly in the displayed example. At more positive electrode potentials the band is shifted to higher wavenumbers, indicating changes in the internal CO bond. The understanding of this chemisorption is of fundamental importance in fuel cell and sensor research.

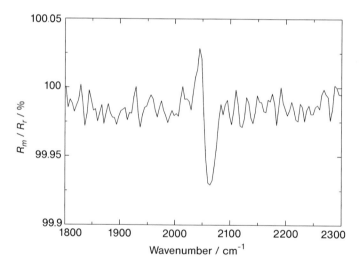

Fig. 5.4 SNIFTIR[3] spectrum of CO_{ad} formed from methanol on a platinum electrode, aqueous solution of 1 M $HClO_4$+1 M methanol, $E_{r,RHE}=0.05$ V, $E_{m,RHE}=0.45$ V, spectrometer BioRad FTS 40.

---

3) SNIFTIR(S) = Subtractively Normalized Fourier Transform Infrared Interfacial Reflection Spectroscopy.

Literature

R. Holze: Surface and Interface Analysis: An Electrochemists Toolbox, Springer Verlag, Berlin 2009.

Spectroelectrochemistry, (R. J. Gale, ed.), Plenum Press, New York 1988.

Electrochemical Interfaces (H. D. Abruna, ed.), VCH, New York 1991.

R. Holze and W. Vielstich, Electrochim. Acta, **33** (1988) 1629.

Experiment 5.4: Electrochromism

Task

Electrochromic color changes of Prussian blue are observed. A simple display is constructed and evaluated.

Fundamentals

Electrochemically induced changes of the color of matter (electrochromism) are interesting in fundamental studies of matter; they also have considerable application potentials. Examples have been studied with polyaniline (see Expt. 5.1). Prussian blue[4] $KFe[Fe(CN)_6]$ is a charge transfer complex prepared easily by the reaction of Fe(III)-ions and hexacanoferrate ions according to

$$K^+ + Fe^{3+} + Fe(CN)_6^{4-} \rightarrow KFe[Fe(CN)_6] \tag{5.1}$$

showing an intense blue color. It can be reduced electrochemically yielding a white product also called Berlin white.

$$K^+ + e^- + KFe[Fe(CN)_6] \rightarrow K_2Fe[Fe(CN)_6] \tag{5.2}$$

This process can be reversed, i.e. the compound can be switched electrochemically between the blue and the white form. This electrochromism suggests its use as an active material in a display.

Execution

Chemicals and instruments

Aqueous solution of $Fe(NO_3)_3$ 0.1 **M**

Aqueous solution of $K_3Fe[Fe(CN)_6]$ 0.1 **M**

Aqueous solution of KNO_3 1 **M**

Stainless steel plate 40×50 mm, approx. 0.5 mm thick[5]

Stainless steel wire (counter electrode)

Petri dish

Fine-grain abrasive paper

Voltage source

---

[4] This material is stable towards light and is used as a blue pigment; other names are Turnbulls blue, Prussian blue, China blue, Paris blue, Milori blues, Toning blue etc.

[5] Measures are approximate; they can be varied widely depending on available material.

Setup

A copper wire is attached to the steel plate. This can be done by, e.g., drilling a hole at a corner of the plate, cutting a thread, and attaching the wire with a screw. To avoid local corrosion, the connection has to be coated with hot melt or epoxy resin. The steel plate is polished with fine abrasive paper.

From 2 ml each of the iron salt solution an ink is prepare; because of its brownish-yellow color it is also called Berlin yellow. A few drops are applied to the freshly polished steel plate; if desirable letters, symbols etc. can be drawn. Upon contact and further exposure the surface a layer of Berlin blue is formed visible as a blue coloration at the bottom of the deposited liquid. After about three minutes the liquid is removed, and the plate is rinsed with distilled water. It is placed in the petri dish, and a loop formed from the stainless steel wire is placed on the bottom of the dish (close to the edge of the dish) as a second electrode.

Procedure

The voltage source is connected to the two pieces of steel. Initially the coating is blue. The steel plate is connected to the minus pole, and the wire to the plus pole. The voltage is increased from zero to about $U=1.4$ V; higher voltages should be avoided because they might result in destruction of the electrochromic material. When the material has changed its color the polarity is reversed, and the blue colour reappears.

Evaluation

On the freshly polished steel surface (traces of the protective chromium oxide layer have just been removed) iron atoms are oxidized by Berlin yellow according to

$$Fe + 2\ Fe[Fe(CN)_6] \rightarrow Fe^{2+} + 2\ KFe[Fe(CN)_6] \tag{5.2}$$

yielding Berlin blue. In the first electrochemical step the Berlin blue is reduced at the cathodically polarized steel plate according to

$$K^+ + e^- + KFe[Fe(CN)_6] \rightarrow K_2Fe[Fe(CN)_6] \tag{5.3}$$

The reaction product is hardly visible. At the steel wire anode water is decomposed yielding dioxygen and hydroxide ions; the amount of oxygen is too small to become visible. After polarity reversal reoxidation proceeds

$$K_2Fe[Fe(CN)_6] \rightarrow K^+ + e^- + KFe[Fe(CN)_6] \tag{5.4}$$

As demonstrated with Mössbauer spectroscopy only the iron ion not co-ordinated with cyanide ions changes its state of oxidation (K. Itaya, T. Ataka, S. Toshima, and T. Shinohara, J. Phys. Chem. **86** (1982) 2415). At the wire electrode hydrogen is evolved by water decomposition.

Literature

M. Nishan, J. Freienberg, and G. Wittstock, Chemkon **14** (2007) 189.
P.M.S. Monk, R.J. Mortimer, and D.R. Rosseinsky: Electrochromism: Fundamentals and Applications, VCH, Weinheim 1995.

6
Electrochemical Energy Conversion and Storage

The general importance of systems for electrochemical energy storage and conversion can hardly be overestimated. Primary batteries of several types and sizes are omnipresent in mobile applications, in hearing aids, toys, pacemakers, music players, and flash lamps. The list is never ending. Secondary, rechargeable systems prevail in mobile phones, laptop computers, measuring instruments, and other mobile applications. Large systems are used in electric vehicles and uninterrupted power supplies and emergency backup systems. Future applications will only grow in numbers and variety. Quite in contrast, the laboratory experiments suitable to demonstrate typical measurements as performed in research and development beyond simply reproducing the manufacturer's data sheet are small in number. In addition, for many conceivable experiments components from actual cells would be required which are hard to obtain. Numerous conceivable experiments involving, e.g., complete charge and discharge cycles are very time-consuming, and may conflict with the standard laboratory schedule. Thus, the following experiments suggest compromises.

Experiment 6.1: Lead Acid Accumulator [1]

Task
The charging and discharging yield of a lead acid accumulator are measured. The current/voltage curves of the lead and the lead dioxide electrodes are measured.

Fundamentals
The measurements of the parameters of the actual electron transfer step in an electrochemical reaction at an electrode (exchange current density j_0 and symmetry factor a) are central tasks in electrochemistry. Besides calculations of these parameters from results of quasi-stationary (see exp. 3.14) and nonsta-

---

1) The terminology is variable and confusing. Although a lead acid battery used as power source for starting the internal combustion engine is a secondary system – and thus an accumulator – it is seldom referred to in this way, except by technological experts. This confusion prevails generally – up to the term "secondary battery".

Experimental Electrochemistry. A Laboratory Textbook. Rudolf Holze
Copyright © 2009 WILEY-VCH Verlag GmbH & Co. KGaA, Weinheim
ISBN: 978-3-527-31098-2

tionary experiments recording current/potential curves under conditions where charge transfer is current limiting is a valuable and frequently employed approach. A simplified evaluation making use of Tafel plots allows the direct determination of both parameters. Experimental requirements for the applicability of the approximation are met at very small current densities where no mass transport limitation is expected. Other conceivable hindrances (i.e. overpotentials) must be excluded by proper definition of the experimental conditions. Small current densities can be easily obtained by using electrodes of high specific surface area: porous electrodes with a high ratio of true (internal) surface area to apparent surface area. Only the former surface is entered into the approximated evaluation.

In the experiment described here, porous electrodes as employed in a car battery are used. Both the porous lead dioxide and the sponge-like lead electrode are highly porous. The use of concentrated sulfuric acid eliminates all other conceivable overpotentials. As well as the acquisition of current/potential curves, further data to be studied in this experiment are easily accessible. Processes in a lead-acid accumulator are shown schematically in Fig. 6.1 (see also EC:441).

Execution
Chemicals and instruments
Lead electrode[2]
Lead dioxide electrode
Aqueous solution of sulfuric acid battery grade (36 wt%, density 1.25 $g \cdot cm^{-3}$)
Hydrogen reference electrode
Cell
3 Multimeters
Power supply
Shunt resistor
X-t-recorder

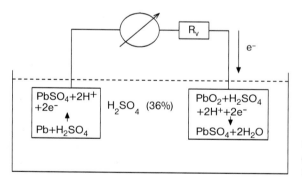

Fig. 6.1 Scheme of processes in a lead-acid accumulator during discharge.

---

[2] Both electrodes can be cut carefully from the respective plates of a dry, precharged lead acid accumulator as sold in car repair shop. Because of the lead content, careful handling is required.

Setup

The demonstration cell used here is a rectangular glass vessel with perpendicular grooves in opposite walls. Into these grooves the electrodes of about 3×5 cm^{-2} size are gently pushed. Electrical contact is made via clamps to the grid in both electrodes. The electrical setup is shown below (Fig. 6.2).

Both electrodes are of the dry-precharged type, i.e. upon adding the sulfuric acid and after the acid has soaked the porous electrodes completely the battery is charged and ready. This will take a few hours. As a reference electrode a hydrogen electrode filled with battery acid is used. If charging with hydrogen is needed this can be easily accomplished by connecting the minus pole of the power supply temporarily to the hydrogen electrode and its plus pole to the lead dioxide electrode. As described in Chapter 1, half the volume of the electrode should be filled with hydrogen. The arrangement of the electrode in the cell is shown in Fig. 6.3.

Procedure

The power supply is used as a constant current source; a voltage of about 20 V is set first with the shunt resistor set to a medium value. After connecting to the cell, the current is adjusted at the power supply to the desired value. For discharge, connectors are reversed.

To verify full charge initially, charging is applied until gas evolution starts. Now the connectors are reversed and the cell is discharged at a current

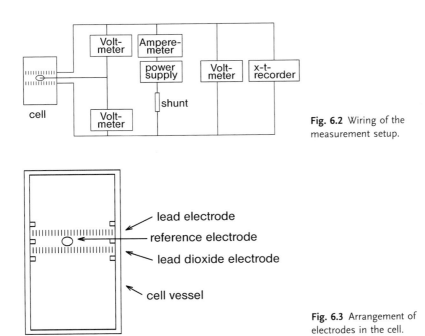

Fig. 6.2 Wiring of the measurement setup.

Fig. 6.3 Arrangement of electrodes in the cell.

$0.3 < I < 0.5$ A[3]) down to the lower voltage limit of 1.7 V or until the discharge curve drops (see Fig. 6.6); this may last somewhere between 0.1 and 1 h. Subsequently charging until gas evolution starts is performed again. If the time needed for charging is shorter than the preceding discharge time no *perpetuum mobile* is present, instead most likely active mass has been lost from an electrode. In this case the cycle has to be repeated. For the second part of the experiment the electrodes must be recharged.

Evaluation

The charge-discharge yield (Faradaic yield) η_{Ah} is calculated according to

$$\eta_{Ah} = \frac{\text{discharge current} \cdot \text{discharge time}}{\text{charge current} \cdot \text{charge time}} \tag{6.1}$$

The energy yield (energy yield) η_{Wh} is calculated

$$\eta_{Wh} = \frac{\int_{t=0}^{\text{discharge time}} I \cdot U(t)\,dt}{\int_{t=0}^{\text{charge time}} I \cdot U(t)\,dt} \tag{6.2}$$

The integrals can be obtained by obtained by determining the areas under the respective curves (e.g., by cutting and weighing the recording paper).

To obtain the current-potential curves (for setup see Fig. 6.2), after measuring the rest potentials ($\eta = 0$ V) the current is increased stepwise with the power supply connected for charging in 50 mA increments until a current of about 500 mA is reached. Readings are taken after the potentials have settled (about 1 min). After reversing polarity the procedure is repeated to obtain the remaining values. Potentials are measured with two multimeters set as voltmeters.

A typical plot of cell voltage as a function of applied current is shown in Fig. 6.4. Figure 6.5 shows the current-potential curves for both electrodes. From the data in Fig. 6.6 the plots shown in Fig. 6.5 could be constructed. This may serve as a way to control the results for consistency.

Figure 6.6 shows charge and discharge curves; from these curves the yields can be obtained. η_{Ah} is calculated fairly simply because a constant current was applied. At the current $I = 0.5$ A applied here $\eta_{Ah} = 0.71$. The energy yield is considerably poorer because of the rather high applied current $\eta_{Wh} = 0.49$.

Analysis of the obtained current-potential curves requires precise knowledge of the actual, true surface areas and well-defined electrode processes.

---

3) The actual value should be set according to the properties of the electrodes. With small or already used electrodes lower values are recommended, with fresh and large electrodes the higher ones.

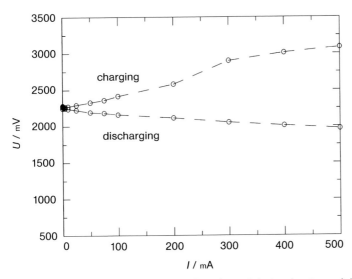

Fig. 6.4 Cell voltage vs. applied current plots obtained during charging and discharging.

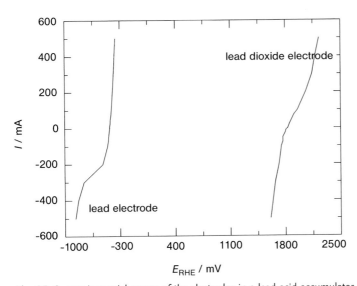

Fig. 6.5 Current/potential curves of the electrodes in a lead acid accumulator.

In the present example lead formation certainly proceeds at the lead dioxide electrode during discharging and lead oxidation is the anodic process at the lead electrode. During charging, hydrogen evolution at the lead electrode and dioxygen evolution at the lead dioxide electrode may compete with the desired charging reactions. Thus, further analysis of these results is not recommended, and instead Expt. 3.14 is suggested.

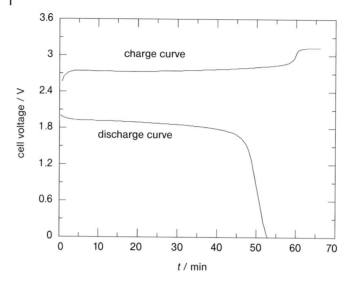

Fig. 6.6 Charge and discharge curves of a lead acid accumulator at $I = 0.5$ A.

Literature

Ullman's Encyclopedia of Industrial Chemistry, VCH, Weinheim **5** 1989, p. 364.
L. F. Trueb and P. Rüetschi, Batterien und Akkumulatoren, Springer-Verlag, Berlin and Heidelberg 1998.
H. Bode: Lead acid batteries, Wiley Interscience, New York 1977.

Questions

- What causes self discharge of a lead acid accumulator?
- What happens during gassing?
- What is sulfation?

Experiment 6.2: Discharge Behavior of Nickel-Cadmium Accumulators

Task

With commercial nickel-cadmium accumulators, discharge curves are recorded at different currents.[4]

Fundamentals

The capacity of a secondary cell (accumulator) is limited by the amount of active mass contained therein. Depending on cell construction and operating conditions the amount of charge (i.e. energy) that can actually be recovered may vary

---

4) Highly interesting temperature effects may become visible when the cell is operated at emperatures well below 0 °C, especially at high discharge currents.

considerably, and this also applies to the delivered cell power. Taking into account the fundamentals of electrode kinetics and the properties of electrolyte solutions and electrical conductors, some general statements are possible. At lower temperatures the capacity of a cell falls. The electrolytic conductance lowered according to the temperature causes a higher internal cell resistance, and this causes a drop in cell voltage, particularly pronounced at higher currents. Thus the final discharge voltage limit is reached earlier. This is the most common reason for the difficulties encountered when starting internal combustion engines on cold winter mornings. Higher discharge currents also result in lower delivered energy values. At higher currents the correspondingly higher electrode overpotentials result in smaller cell voltage, and again the end of the discharge is reached earlier. In addition, active masses may not be converted completely. Reaction products (lead sulfate) deposited in unfavorable positions (e.g., closing pores) and local depletion of reactants (sulfuric acid) are major reasons. This relationship is illustrated in Ragone diagrams (EC:461).

Experimental determination of available capacity of an accumulator can be done in two ways. Discharge can be performed at constant current, which requires a more elaborate experimental setup (a current source), but the result is obtained most easily just by multiplying the adjusted current by the passed time. Experimentally more simple is the second approach: discharge across a constant (shunt) resistor. Correlated with the decreasing cell voltage the flowing current also decreases, and thus determination of the discharge capacity becomes more complicated. In both cases the cell voltage must be monitored to avoid deep discharge or even cell voltage reversal resulting in damage to the cell up to explosion. A careful examination of the effect of operating temperature on the discharge capacity is even more complicated. A careful temperature control requires a sufficiently powerful cryostat. Self-heating of the cell caused by Joule heating of conductors in the cell may become considerable sources of heat, in particular at high discharge currents.

Execution
Chemicals and instruments
Rechargeable nickel-cadmium accumulator of size AAA (micro cell)
Adjustable current source
Shunt resistor (10 Ω)
Cell voltage supervisor circuit (see Appendix)
X-t-recorder

Setup
The freshly charged cell (according to manufacturers recommendations, this requires in most cases charging at a current equivalent to 1/10 of the nominal cell capacity for 14 h) is connected via the shunt resistor (this helps to avoid damage to the current source) and the supervisor circuit to the power supply, the x-t-recorder is connected to the cell terminals.

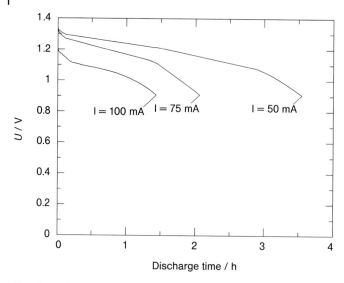

Fig. 6.7 Discharge curves of a nickel-cadmium accumulator at different discharge currents.

Procedure

Recording the cell voltage vs. time plot is started when the discharge current is applied to the cell until a preset cell voltage is reached.

Evaluation

Figure 6.7 shows a typical set of discharge curves obtained with a nickel-cadmium accumulator of size AAA and a nominal capacity of 250 mAh stated by the manufacturer. At $I=100$ mA a capacity of 143 mAh, at $I=75$ mA a capacity of 154 mAh, and at $I=50$ mA a capacity of 170 mAh were obtained.

Question

Is there a thermodynamic explanation of the observations?

Experiment 6.3: Performance Data of a Fuel Cell

Task
Current-voltage data of a hydrogen-dioxygen fuel cell are measured under typical operating conditions.

Fundamentals
A fuel cell is an electrochemical energy converter. A fuel (e.g., hydrogen) reacts inside with an oxidant (e.g., dioxygen) converting chemical into electrical energy (and heat). In contrast to primary and secondary cells the active materials are not stored in the cell. The cell contains only the electrodes needed for the electrochemical reactions, the electrolyte (or the electrolyte solution fixed in some suitable material), and further components needed for operation.

As already discussed, with the lead acid accumulator porous electrodes with large real surface areas are required. These may be sintered metal powders, Raney metals or polymer-blended active carbons. In addition, in many cases further catalysts are deposited on the porous material to accelerate the desired electrode reaction. When a gas is used as reactant instead of a two-phase boundary (solid/liquid; electrolyte solution with liquid reactant and solid electrode) a three-phase boundary (solid/liquid/gas) is established and must be maintained during operation. This requires particular features like, e.g., different scales of porosity or hydrophobization.

The schematic setup of a fuel cell is shown in Fig. 6.8.

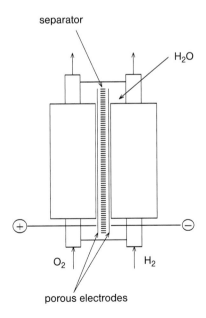

Fig. 6.8 Schematic setup of a hydrogen-dioxygen fuel cell (simplified cross section).

The cell reaction is

$$2 H_2 \text{ (gas)} + O_2 \text{ (gas)} \rightarrow H_2O \text{ (liquid)} \qquad (6.3)$$

Hydrogen diffuses into the porous anode, is dissolved in a thin electrolyte film covering the catalyst surface, and diffuses to active sites where oxidation proceeds. The last diffusion step is particularly limiting because of the low solubility and diffusion coefficient of hydrogen in water. This hindrance is minimized by keeping the electrolyte film thin and the true active surface area large to maintain a small true current density. Protons resulting from the oxidation move across the electrolyte. On the other side they react with the products of dioxygen reduction, which proceeds in the same manner as hydrogen oxidation. The formed reaction water must be removed from the cell to avoid undesirable dilution effects. Numerous types of construction of fuel cells, with different electrolytes, operating temperature, cell construction etc., have been developed and described in the abundant literature. Instead of a separator keeping the liquid electrolyte solution in place as shown in Fig. 6.8, solid polymer electrolytes based on ion-exchanger membranes have been proposed. This results in a significantly simplified cell construction, and in addition the fairly thin electrolyte film (<0.5 mm) causes a very small internal cell resistance. Thus, significant increases of power density without additional losses are possible.

Execution
Chemicals and instruments
Hydrogen gas
Dioxygen gas
Solid poymer electrolyte fuel cell
Voltmeter
Ammeter

Setup
The fuel cell is connected to the hydrogen source according to manufacturers suggestions. To obtain current/cell voltage data a voltmeter is connected to the cell. Current is measured with the ammeter and adjusted with a shunt resistor serving as a load. Instead, a shunt combined with a current source may be used; this simplifies current adjustment considerably. Operation with pure dioxygen instead of air may require a special gas feed at the cathode.

Procedure
Starting at low currents, current-cell voltage data are obtained. After adjusting the desired load (current), in most cases a few minutes are needed until data have settled to stable values. Manufacturers' suggestions regarding operating conditions, in particular maximum current and temperature, must be observed. Polarity reversal must be carefully avoided; it will most likely result in cell destruction.

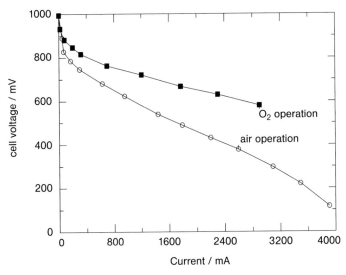

Fig. 6.9 Current-cell voltage data of a commercial solid polymer electrolyte fuel cell operated with hydrogen and air or
dioxygen.

Evaluation

A typical set of results obtained with a commercial cell (active electrode area
12 cm^2) is shown in Fig. 6.9.

The cell voltage at zero current (U_0) is independent of feed gas. As expected,
the value calculated from thermodynamic data $U_0 = 1.229$ V is approached. Under load, the performance of the cell fed with dioxygen is superior. Formation
of diffusion hindrances (dioxygen has not to diffuse through layers of inert gas,
the enrichment of inert gas after some time of operation even aggrieving this
hindrance) is avoided.

Literature

K. Kordesch and G. Simader: Fuel Cells and their Applications, VCH, Weinheim 1996.

7
Electrochemical Production

Electrochemical methods of production are of considerable importance in various branches of industry; on average they consume about 7% of the electicity generated. They are important in metal winning and refining, production of numerous inorganic and organic chemicals, and in machining and surface treatment of materials and components. Typical examples from inorganic chemistry are the chlor-alkaline electrolysis or the winning of aluminum or copper by electrolysis of molten salts or electrolyte solutions. In organic chemistry, but possible applications are much more numerous, their overall volume is nevertheless far smaller than in inorganic chemistry. Procedures are developed to modify functional groups or to form new C-C bonds by reduction or oxidation reactions. Typical examples surface treatment, galvanic coating with copper, gold, lead, and electrochemical machining.

Industrial cells are complicated, large, and, because of sometimes extreme operating parameters (temperature, currents), hard to model in the laboratory. Some salient features can nevertheless be studied on a laboratory scale except for electrochemical machining.

Experiment 7.1: Cementation Reaction

Task
Copper deposition from aqueous solution on iron scrap and with zinc dust is examined.

Fundamentals
A less noble metal added in elementary form to a solution of a more noble one will result in precipitation of the noble one, and a corresponding amount of the less noble one is dissolved. This known as a cementation reaction. The reaction of copper ions with iron is given as:

$$Cu^{2-} + Fe \rightarrow Cu + Fe^{2-} \tag{7.1}$$

This reaction is technologically important in winning silver, copper, and other metals. In copper winning it has been replaced by electrolytic extraction. This

Experimental Electrochemistry. A Laboratory Textbook. Rudolf Holze
Copyright © 2009 WILEY-VCH Verlag GmbH & Co. KGaA, Weinheim
ISBN: 978-3-527-31098-2

process can also be observed in corrosion (see above). Finally, it can be used as a simple procedure to produce metal coatings (e.g., copper on iron).

Execution
Chemicals and instruments
Aqueous solution of $CuSO_4$, approx. 1 **M**
Iron powder
Zinc powder
Small beaker

Setup
The copper sulfate solution is placed in the beaker; for better visibility the beaker is placed on white paper.

Procedure
To the solution some iron powder is added. When these reactants are slowly stirred decoloration occurs. If necessary, further iron powder should be added. The procedure is repeated with zinc powder.

Evaluation
Decoloration indicates reduction of copper ions, which is also apparent from the slight reddish color of the iron caused by the copper coating. Because the reaction is a heterogeneous one (it proceeds only at the interface iron particle/ copper ion solution), in the case of coarse iron particles a larger amount is necessary because finer particles have a larger specific surface area. With zinc dust the reaction proceeds particularly fast. The cementation reaction occurs also in the Kipp apparatus for production of hydrogen, when some copper sulfate is added to the granulated zinc and the sulfuric acid. By copper deposition local elements composed of copper and zinc are formed. Subsequently on the copper surface hydrogen evolution by proton reduction proceeds very effectively, whereas at the zinc surface anodic dissolution occurs. Because of the fairly high hydrogen overvoltage on zinc, hydrogen evolution on the zinc is less rapid.

Experiment 7.2: Galvanic Copper Deposition

Task
Current distribution during copper deposition at a split copper electrode is studied to understand throwing power.

Fundamentals
In electrochemical processes at the phase boundary solution/electrode it is generally assumed that the rate of reaction (conversion of matter etc.) is the same everywhere, i.e. local current density is the same at all locations. In many cases

and when favorable experimental conditions are established (highly conducting electrolyte solutions, low current density, symmetric arrangement of working, and counter electrode) this will be approximately true. In practical metal deposition (and other electrochemical processes) these conditions are frequently not met. Complicated forms and surfaces of components, poorly conducting solutions, and highly asymmetric arrangement of electrodes may result in highly unequal current distribution and consequently different rates of local metal deposition etc. This can have important, even devastating practical consequences. A result of unequal current distribution is that at places of low current density only thin metal layers are formed. When the properties of a component (corrosion stability, hardness, wear resistance) depend on this thickness undesirable local variations of these properties may be the consequence.

Various possibilities are available to prevent these consequences. Suitable construction of the component, favorable arrangement of the counterelectrode(s), pumped or stirred highly conducting electrolyte solutions are a few of them. The capability of a given solution to suppress uneven metal deposition is called "throwing power"; a high throwing power corresponds to good distribution even in less favorable cases. This property can easily be investigated with a split electrode enabling separate measurements of the current flowing into different electrode segments. Assuming that the electrochemical properties of the copper plate employed in this experiment are the same over the whole surface and that copper ion concentration is isotropic in this experiment the primary current distribution is studied. When apart from this current distribution, which is only controlled by cell and electrode geometry, locally different surface properties of the electrodes (like, e.g., local hindrance of copper deposition by deposits of contaminants) are taken into account, the secondary current distribution is studied. Taking into account finally local concentration differences, the result is the ternary current distribution.

Execution
Chemicals and instruments
Split copper electrode
Copper wire
Aqueous solution of $CuSO_4$, approx 0.1 **M**
2 Identical ammeters
Adjustable power source
Beaker

Setup
Two sheets of copper foil of equal size (e.g. 3×4 cm) are cut. At the shorter edge near the middle a copper wire is soft soldered on. After insulating the solder points with, e.g., adhesive tape or hot melt resin the sheets are fixed together with a sheet of plastic foil as insulator between them. Edges are coated with adhesive tape. The wires are conneceted via an ammeter to the minus pole

of the power source. The plus pole is connected to a copper wire dipping in the solution. At both ammeters the same current range is set.

Procedure

The total current is set at the power source, and partial currents into both electrodes are read at the ammeters. The experiment may be done at different copper ion concentrations, cell geometries, placement of the copper wire anode, use of an additional anode, electrode distance, stirring etc.

Evaluation

In a typical experiment with a dilute copper sulfate solution and only one anode placed in front of one sheet at a total current of $I=50$ mA the current into the electrode in front of the anode was $I_f = 34.6$ mA, whereas the current to the back electrode was only $I_b = 15.4$ mA. At $I=100$ mA the distribution was $I_f = 67$ mA and $I_b = 33$ mA.

Experiment 7.3: Electrochemical Oxidation of Aluminum

Task

By anodic oxidation of an aluminum surface an oxide film is generated, which is subsequently colored and sealed.

Fundamentals

On practically all surfaces of non-noble metals thin oxide films are formed once the metal is exposed to air. This oxide layer can be a protective, passivating one if it is mechanically and chemically stable. The coating of Al_2O_3 formed most easily on aluminum is a particularly good example. Naturally formed layers are insufficient for technical application. By electrochemical oxidation in a suitable electrolyte solution the layer can be grown and improved (up to about 0.02 mm) [1].

Execution
Chemicals and instruments
Sheet of aluminum plate, about 2×4 cm
Aluminum wire for hanging the sheet
Strip of aluminum sheet, approx. 1×10 cm
Degreasing agent (organic solvent)
Aqueous solution of 1 **M** NaOH
Aqueous solution of 0.2 **M** HNO_3
Aqueous solution of 2 **M** H_2SO_4

---

1) In German this process is called "Eloxal-Verfahren".

Aqueous solution of 0.1 **M** $(NH_4)_3Fe(C_2O_4)_3$ [2]
Aqueous solution of ammonium acetate 0.1 wt% [3]
Power supply
Beaker

Setup
For anodic oxidation the aluminum sheet is suspended by the wire in the beaker filled with sulfuric acid. The aluminum strip is wound into a coil surrounding the sheet and also is suspended in the solution. Both electrodes are connected to the power supply.

Procedure
The aluminum sheet is carefully degreased, rinsed with plenty of water, and etched in the sodium hydroxide solution for a few minutes. Vigorous hydrogen evolution will be observed. Subsequently the sheet is neutralized in the nitric acid solution and rinsed with water again. Electrolytic oxide layer formation is performed in the sulfuric acid solution at $T=26\,°C$ at $j=10...20\ mA\cdot cm^{-2}$ for about 15 min, the applied voltage should not exceed 15 V. For intense coloring a thick oxide layer is desirable, and in this case layer formation should go on for 30 min. The sheet is taken out of the acid solution, rinsed and put into the warm $(T=70\,°C)$ coloration solution (any water soluble dye may be tried) for a few minutes. When the desired depth of coloration is achieved the sheet is removed and rinsed. If needed coloration may be commenced again. Final sealing is achieved by dipping the sheet in boiling water for 15...30 min, better in a boiling solution of 5 g nickel acetate and 5 g boric acid/l water for layers without coloration or with a golden color. For other colors the ammonium acetate solution should be used for about 30 min.

Evaluation
Treatment of the aluminum sheet with caustic resulted in a uniform, well-defined surface yielding a regular oxide film during subsequent oxidation. Aluminum alloy containing silicon shows unfavorable properties at this point because of the different properties of silicon and aluminum. This will certainly become obvious in the form of faint and uneven coloration. During the final sealing the aluminum oxide is partially hydrated. This results in volume increase, densification of the layer, and enhanced mechanical stability.

---

2) The necessary solution of iron oxalate can be prepared easily by mixing respective amounts of solutions of iron(III) chloride and ammonium oxalate.

3) This solution will be needed only for sealing the surface layer; an alternative is given below.

Experiment 7.4: Kolbe Electrolysis of Acetic Acid

Task
By Kolbe electrolysis of acetic acid, ethane is produced.

Fundamentals
As early as 1833 M. Faraday reported the generation of carbon dioxide and a hydrocarbon during electrolysis of a solution of potassium acetate. He initially assumed that these were secondary products of the attack of anodically generated dioxygen on the acetate. To his surprise a reduced compound with carbon atom in a lower state of oxidation – ethane – was formed at the anode. In 1848–1850 H. Kolbe studied the reaction later named after him more intensely. He proposed in general terms that upon loss of a unit of carbon dioxide from the anodically discharged carboxylic acid anion a radical is formed which subsequently produces a hydrocarbon. The reaction of acetic acid studied here is thus

$$2 \ CH_3COO^- \rightarrow 2 \ CO_2 + C_2H_6 + 2e^- \tag{7.2}$$

According to the current state of knowledge, free radicals are assumed as intermediates:

$$CH_3COO^- \rightarrow CH_3^\bullet + CO_2 \tag{7.3}$$

Because as an anode a platinum electrode of small surface area is used and a large current (i.e. a large current density) is applied, the stationary local concentration of free radicals is substantial. Recombination of the radicals yields the hydrocarbon ethane:

$$2 \ CH_3^\bullet \rightarrow C_2H_6 \tag{7.4}$$

The presence of a radical intermediate can be demonstrated by, e.g., a trapping reaction with styrene. In this case polymerization of styrene (a radical reaction) is initiated.

Execution
Chemicals and instruments
0.61 mol Sodium acetate hydrate (50 g)
0.87 mol Glacial acetic acid (50 ml)
Platinum wire electrode (anode)
Copper wire electrode (cathode)
Power supply
Crystallization or petri dish
Buret (with two valves, 50...100 ml volume)
Beaker

Setup
For the electrolysis the setup shown in Fig. 7.1 can be used.

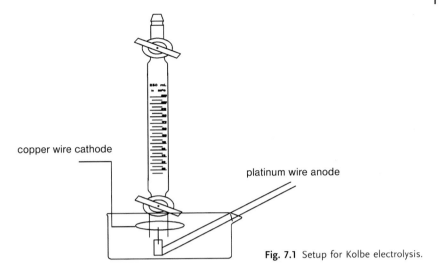

copper wire cathode

platinum wire anode

Fig. 7.1 Setup for Kolbe electrolysis.

Procedure

The electrolyte solution prepared from the two components and water is put into the dish. The buret is mounted above the dish with its lower end extending into the solution. The platinum wire is fixed with its tip extending into the lower end of the buret. The copper wire is placed around the lower end above its bottom edge. In this way no cathodically generated hydrogen can enter the buret. By suction the buret is filled up to the top valve with solution. An electrolysis voltage of about $U=12$ V results in an appreciable rate of gas evolution.

Because carbon dioxide has a non-negligible solubility in the electrolyte solution a first electrolysis should be performed until the buret is filled with gas. Now the solution saturated with gaseous products is sucked again into the buret. Electrolysis is performed a second time until the volume between the two valves is filled with gas. After closing the valves the gas can be analyzed. If it is extracted with an aqueous solution of NaOH, reduction of the gas volume indicates reaction of the initially formed CO_2 yielding dissolved carbonate. When the residual gas is carefully released and ignited it burns slowly with a bluish flame typical of a burning hydrocarbon.

Experiment 7.5: Electrolysis of Acetyl Acetone

Task

By indirect electrolysis of acetyl acetone (pentane-2,4-dione) in a nonaqueous electrolyte solution 3,4-diacetylhexan-2,5-dione is produced.

Fundamentals

Indirect anodic oxidation of acetyl acetone yields an iodine substituted intermediate, which reacts in a coupling reaction to form 3,4-diacetyl-hexane-2,5-
dione. Figure 7.2 shows the overall transformation.

Closer inspection of the reaction pathway reveals further details of this anodic
dimerization. According to the reaction scheme in Fig. 7.2 the acidic hydrogen
atom in position 3 can be split off yielding a proton and the acetyl acetone anion. At the anode, iodide is oxidized to iodine. Iodine reacts with this anion
forming an iodine-substituted (in 3-position) acetyl acetone. Because this iodine
substituent is fairly reactive, further substitution by another acetyl acetone anion
at the respective carbon atom in the 3-position proceeds smoothly. Upon release
of an iodide ion the product is formed. This sequence is shown in Fig. 7.3.

Execution
Chemicals and instruments
250 ml Acetone
40 mmol Acetyl acetone (4 g)
3.3 mmol NaI (0.5 g)
Platinum wire net electrode

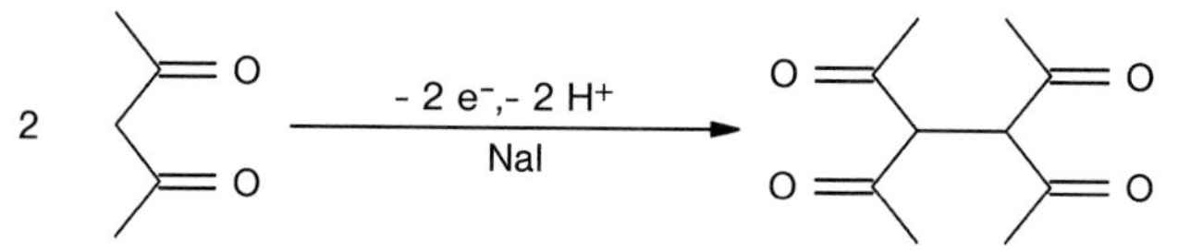

Fig. 7.2 Reaction scheme of the formation of 3,4-diacetyl-hexane-2,5-dione.

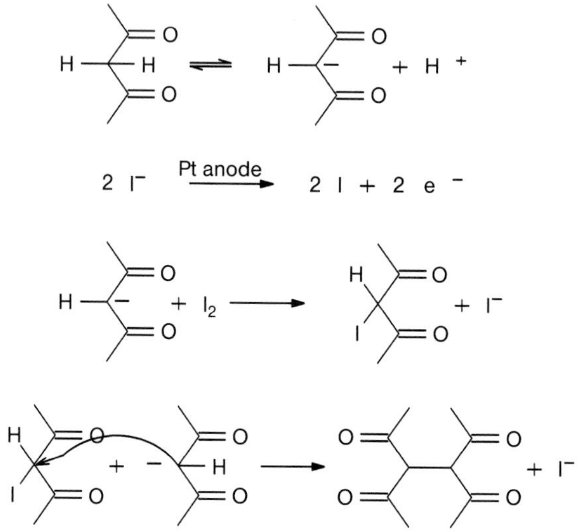

Fig. 7.3 Mechanism of the anodic dimerization of acetyl acetone.

Iron wire cathode
Beaker
Magnetic stirrer plate
Magnetic stirrer bar
Power supply (90 V, 0.5 A)

Setup

The iron wire electrode wound into a spiral is placed as the cathode in the center of the beaker, and the platinum wire net as anode is placed around it (see Fig. 7.4). The solution of acetyl acetone in acetone is poured into the beaker. The height of the electrodes is adjusted to avoid a short circuit between them and collision with the stirrer bar.

Procedure

A DC voltage of about 60 V is applied to the electrodes[4]. Initially only a very small current or no current at all will flow. Upon addition of NaI the current grows rapidly. It is limited to $I=0.5$ A at the power supply. If this current is not reached, further NaI is added. Electrolysis at $I=0.5$ A is performed for two hours. The temperature of the reaction mixture rises because of Joule heating. Evaporating acetone must be replaced. In the case of too rapid acetone evaporation because of too much Joule heating, significant losses of iodine are possible. In this case the current falls rapidly. Addition of some NaI compensates for this loss. After the end of electrolysis (after about two hours or after passage of about twice the stoichiometrically necessary electrical charge) the electrodes are removed. Acetone is evaporated in a rotary evaporator. The brown-colored product is dissolved in 10 ml acetone and placed in a freezer overnight. The pre-

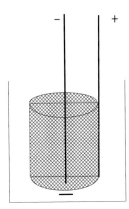

Fig. 7.4 Setup for electrolysis of acetyl acetone.

4) According to safety regulations applicable to electrical devices and their use, the DC voltage applied here requires more care than the voltages encountered in other experiments. This extended care includes precautions against short circuits and accidental touching of electric components.

cipitated crystals are collected on a Buchner funnel. Washing with water results in almost complete decoloration (washing with a mixture of water/acetone (5/1 vol.) could easily result in complete dissolution of the product). The obtained crystals recrystallize at about 160 °C; at exactly 160 °C first sublimation is observed. The white crystals melt at 192…194 °C, in agreement with literature data. The raw yield in the experiment described here is 12%.

Literature

M. N. Elinson, T. L. Lizunova, and T. I. Nikishin, Bull. Acad. Sci. USSR (Engl. Transl.) **41** (1992) 123.

Experiment 7.6: Anodic Oxidation of Malonic Acid Diethylester

Task

By indirect oxidation of malonic acid diethylester in a nonaqueous electrolyte solution, 2,3-bis-ethoxycarbonyl-succinic acid diethyl ester is synthesized.

Fundamentals

Anodic oxidation of malonic acid diethylester in the presence of iodide yields an iodine-substituted intermediate, which forms 2,3-bis-ethoxycarbonyl-succinic acid diethyl ester in a coupling reaction. Figure 7.5 shows the overall reaction equation. The reaction mechanism is the same as previously discussed for the electrolysis of acetyl acetone (see preceding experiment).

Execution
Chemicals and instruments
250 ml Acetone
40 mmol Malonic acid diethylester (6.07 ml)
3.3 mol NaI (0.5 g)
Platinum wire net electrode
Iron wire cathode
Beaker
Magnetic stirrer plate
Magnetic stirrer bar
Power supply (90 V, 0.5 A)

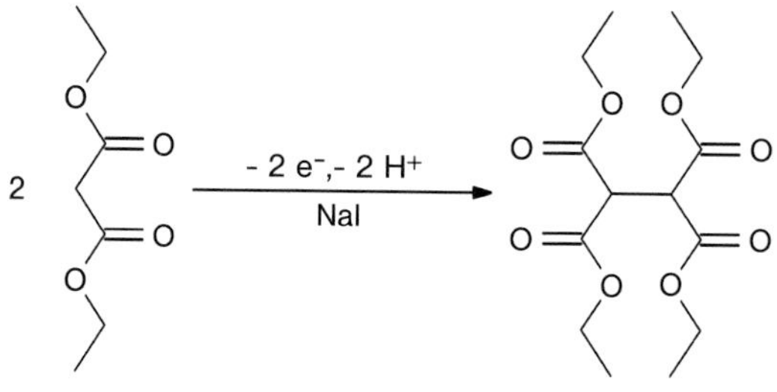

Fig. 7.5 Reaction scheme for the formation of ethentetracarboxylic acid ethylester.

Setup

The iron wire electrode wound into a spiral is placed as cathode in the center of the beaker, the platinum wire net as anode around it (see Fig. 7.4 above). The solution of malonic acid diethylester in acetone is poured into the beaker. The height of the electrodes is adjusted to avoid short circuit between them and collision with the stirrer bar.

Procedure

A DC voltage of about 60 V is applied to the electrodes[5]. Initially only a very small current or no current at all will flow. Upon addition of NaI the current grows rapidly. It is limited to $I = 0.5$ A at the power supply. If this current is not reached, further NaI is added. Electrolysis at $I = 0.5$ A is performed for two hours. The temperature of the reaction mixture rises because of Joule heating. Evaporating acetone must be replaced. In the case of too rapid acetone evaporation because of too much Joule heating, significant losses of iodine are possible. In this case the current falls rapidly. Addition of some NaI compensates for this loss. After the end of electrolysis (after about two hours or after passage of about twice the stoichiometrically necessary electrical charge) the electrodes are removed. Acetone is evaporated in a rotary evaporator. The brown-colored product is dissolved in 10 ml acetone and placed in a freezer overnight. The precipitated crystals are collected on a Büchner funnel. Washing with water results in almost complete decoloration. Further washing with a mixture of water/acetone (5/1 vol.) results in an even lighter color. The obtained crystals recrystallize at about 160 °C; at exactly 160 °C first sublimation is observed. The white crystals melt at 73…74 °C, in agreement with literature data. The yield in the experiment described here is 59%.

Literature

M. N. Elinson, T. L. Lizunova, and T. I. Nikishin, Bull. Acad. Sci. USSR (Engl. Transl.) **37** (1988) 2285.

Experiment 7.7:
Indirect Anodic Dimerization of Acetoacetic Ester (3-oxo-butyric acid ethyl ester)

Task

By indirect oxidation of acetoacetic ester in a nonaqueous electrolyte solution 2,5-dioxo-hexane-dicarboxylic acid-(3.4)-diethyl ester is prepared.

---

5) According to safety regulations applicable to electrical devices and their use, the DC voltage applied here requires more care han the voltages encountered in other experiments. This extended care includes precautions against short circuits and accidental touching of electric components.

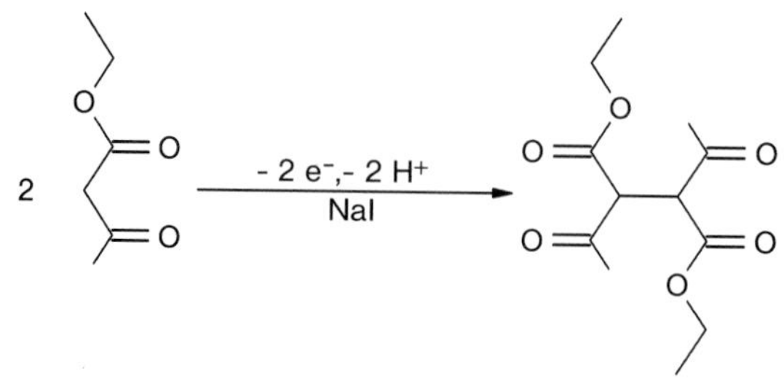

Fig. 7.6 Reaction scheme for the formation of 2,5-dioxo-hexane-di-carboxylic acid-(3.4)-diethyl ester.

Fundamentals

Anodic oxidation of 3-oxo-butyric acid ethyl ester in the presence of iodide yields an iodine-substituted intermediate, which is converted in a coupling reaction into 2,5-dioxo-hexane-dicarboxylic acid-(3.4)-diethyl ester. Figure 7.6 shows the overall reaction equation. The reaction mechanism is the same as previously discussed for the electrolysis of acetyl acetone (see preceding experiments).

Execution

Chemicals and instruments

250 ml Acetone
40 mM 3-Oxo-butyric acid ethyl ester (4 g)
3.3 mM NaI (0.5 g)
Platinum wire net electrode
Iron wire cathode
Beaker
Magnetic stirrer plate
Magnetic stirrer bar
Power supply (90 V, 0.5 A)

Setup

The iron wire electrode wound into a spiral is placed as cathode in the center of the beaker, the platinum wire net as anode around it (see Fig. 7.4 above). The solution of malonic acid diethylester in acetone is poured into the beaker. The height of the electrodes is adjusted to avoid a short circuit between them and collision with the stirrer bar.

Procedure

A DC voltage of about 60 V is applied to the electrodes[6]. Initially only a very small current or no current at all will flow. Upon addition of NaI the current grows rapidly. It is limited to $I=0.5$ A at the power supply. If this current is not

---

[6] According to safety regulations applicable to electrical devices and their use, the DC voltage applied here requires more care han the voltages encountered in other experiments. This extended care includes precautions against short circuits and accidental touching of electric components.

reached, further NaI is added. Electrolysis at $I=0.5$ A is performed for two hours. The temperature of the reaction mixture rises because of Joule heating. Evaporating acetone must be replaced. In the case of too rapid acetone evaporation because of too much Joule heating, significant losses of iodine are possible. In this case the current falls rapidly. Addition of some NaI compensates for this loss. After the end of electrolysis (after about two hours or after passage of about twice the stoichiometrically necessary electrical charge) the electrodes are removed. Acetone is evaporated in a rotary evaporator with only a slight vacuum applied; instead, acetone can also be distilled off carefully in a water bath in order to avoid coevaporation of the product. The brown-colored product is dissolved in 10 ml acetone and placed in a freezer overnight. The precipitated crystals are collected on a Buchner funnel. Washing with a small volume of a mixture of water/acetone (5/1 vol. yields a crystalline product melting at 84 °C in agreement with literature data. The yield in the experiment described here is 20%.

Literature

M. N. Elinson, T. L. Lizunova, and T. I. Nikishin, Bull. Acad. Sci. USSR (Engl. Transl.) **41** (1992) 123.

Experiment 7.8: Electrochemical Bromination of Acetone

Task

By electrochemical bromination of acetone bromoform is synthesized.

Fundamentals

In the classical haloform reaction in organic synthesis, an organic compound with an oxidizable methyl group is treated with a hypohalite in alkaline solution. The halogenized intermediate splits under the influence of the alkaline into the haloform and a carboxylic acid. The reaction is shown simplified in Fig. 7.7.

This process can also be performed electrochemically. The required halogen is formed electrochemically from the respective halogenide. It then reacts with the organic compound. In the example discussed here the simplified anode reaction is:

$$3\ Br^- + 4\ OH^- + (CH_3)_2O \rightarrow CHBr_3 + CH_3COO^- + 3\ H_2O \tag{7.5}$$

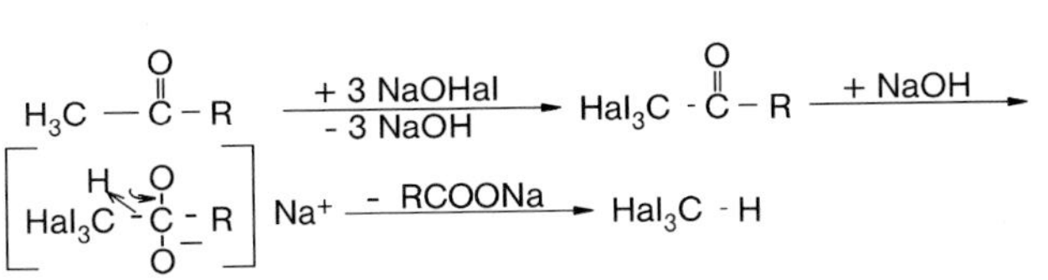

Fig. 7.7 Mechanism of the haloform reaction.

At the cathode, hydrogen evolution proceeds according to

$$6 \ H_2O + 6 \ e^- \rightarrow 3 \ H_2 + 6 \ OH^- \tag{7.6}$$

Because an undivided cell is used, the cathode reaction shifts the pH of the reaction solution into the alkaline region. Although hydroxyl ions are required for the reaction, their excess is undesirable because this could support alkaline disproportionation of the anodically formed bromine according to

$$3 \ Br_2 + 6 \ OH^- \rightarrow 5 \ Br^- + BrO_3^- + 3 \ H_2O \tag{7.7}$$

Under unfavorable conditions this competing reaction may supersede bromoform production. The pH of a hydrogen carbonate solution has turned out to be particularly suitable. The formed hydroxyl ions slowly turn this solution into a carbonate-containing one. In this solution, acetone is directly oxidized into acetic acid and carbon dioxide. To suppress this reaction the solution is saturated with carbon dioxide gas purged into the solution to maintain the concentration of hydrogen carbonate. With high Faradaic yield, bromine-containing bromoform is produced, which fortunately can be purified easily.

In the undivided cell, bromine and bromoform can be reduced at the cathode resulting in a loss of Faradaic yield. Addition of potassium chromate suppresses this reaction strongly. Presumably chromate forms a layer of chromium oxide on the cathode, strongly inhibiting this undesired side reaction. Use of an alcohol instead of acetone (see iodoform synthesis, Expt. 7.9) is impossible because at the high anode potential needed for bromide oxidation to bromine alcohol oxidation may occur.

Execution
Chemicals and instruments
0.15 mol KBr (12.5 g)
5 mmol $KHCO_3$ (7.6 g)
7.5 ml Acetone
0.42 mmol K_2CrO_4 (0.125 g)
Water
2 Platinum electrodes, area about 1 cm^2
Beaker
Carbon dioxide
Power supply (90 V, 0.5 A)

Setup
The ingredients are dissolved in 75 ml water and poured into the beaker. The beaker is put into a cooling bath. The electrodes are inserted and connected to the power supply.

Procedure
Bromoform poses health risks. Electrolysis as well as further handling of the reaction mixture must be performed carefully in a fume hood. Electrolysis is per-

formed at a current density of approx. $j=0.1$ A·cm^{-2}. Higher current densities result in excessive gas evolution and the risk of product loss with the gas stream. The suggested mixture needs about seven hours of electrolysis time. Larger electrodes or smaller amounts of reactants result in correspondingly shorter times of electrolysis. A temperature of about $T=17\,^\circ$C should be maintained during electrolysis. Carbon dioxide is passed through the solution continuously. When a yellow coloring of the solution is observed the stream of gas must be increased. Formation of the bromine-containing product can be observed after a short time of electrolyis, and small brown droplets collect at the bottom of the electrolysis cell. After the end of the electrolysis, this fraction is collected with a separating funnel and purified with a small volume of acetone-containing soda solution. In this example 2 ml bromoform were obtained; this represents a yield of about 9%.

Experiment 7.9: Electrochemical Iodination of Ethanol

Task

Iodoform is synthesized by electrochemical iodination of ethanol.

Fundamentals

Analogously to the electrochemical bromination of acetone the formation of iodoform is possible. Because the anodic electrode potential necessary for iodine formation from iodide is substantially lower than the potential for bromine formation, an organic starting compound more sensitive towards anodic oxidation can be employed. In this case it is ethanol instead of acetone.

At the anode, iodine is formed from the alkaline solution of potassium iodide according to

$$2\ I^- \rightarrow I_2 + 2e^- \tag{7.8}$$

Iodine disproportionates according to

$$3\ I_2 + 4\ OH^- \rightarrow 2\ HIO + IO^- + 3\ I^- + H_2O \tag{7.9}$$

As an undesired follow-up reaction formation of iodate may occur:

$$2\ HIO + IO^- \rightarrow IO_3^- + 2\ H^+ + 3\ I^- \tag{7.10}$$

In the absence of further reaction partners (here: an alcohol), the following overall reaction can be stated:

$$3\ I_2 + 6\ OH^- \rightarrow IO_3^- + 5\ I^- + H_2O \tag{7.11}$$

In the presence of an alcohol as a competing reaction formation of iodoform is possible. With ethanol the sum reaction equation is

$$5\ I_2 + CH_3CH_2OH + 9\ OH^- \rightarrow CHI_3 + CO_3^{2-} + 7\ I^- + 7\ H_2O \tag{7.12}$$

The mechanism is the same as the one discussed for bromoform formation. The product distribution into iodoform and iodate depends on the solution composition. A more alkaline solution favors iodate formation, while a less alkaline one supports iodoform formation. Consequently in this experiment a carbonate-containing solution is employed. Closer inspection of the processes at anode and cathode in the undivided cell reveals a possible problem. At the anode hydroxyl ions are consumed according to [7]

$$10 \; I^- + CH_3CH_2OH + 9 \; OH^- \rightarrow CHI_3 + CO_3^{2-} + 7 \; I^- + 7 \; H_2O + 10 \; e^- \qquad (7.13)$$

When ten mols of electrons are consumed, nine mols of hydroxyl ions are consumed also. At the cathode the following overall equation is valid:

$$10 \; H_2O + 10 \; e^- \rightarrow 10 \; OH^- + 5 \; H_2 \qquad (7.14)$$

Taking both equations together, hydroxyl ions are generated. In an undivided cell this will shift the pH in the alkaline direction. Accordingly, the formation of iodate will be favored. To suppress this, carbon dioxide is sparged into the solution. Because addition of too much CO_2 will shift the pH to values too low for fast iodoform generation, the actual addition of CO_2 should be kept at the minimm amount necessary.

Execution
Chemicals and instruments
12 mmol KI (2.04 g)
15.7 mmol Na_2CO_3 (1.66 g, free of water)
71 mmol Absolute ethanol (4.17 ml)
16.7 ml Water
2 Platinum electrodes, area about 2 cm^2
Beaker
Carbon dioxide gas
Power supply

Setup
The ingredients are mixed and poured into the beaker. The beaker is put into a cooling bath. The electrodes are inserted and connected to the power supply.

Procedure
Iodoform poses health risks. Electrolysis as well as further handling of the reaction mixture must be performed carefully in a fume hood. Electrolysis is performed at a current density of approx. $j=0.1 \; A \cdot cm^{-2}$. Higher current densities result in excessive gas evolution and the risk of product loss with the gas stream. The suggested mixture needs about four hours of electrolysis time. Larger electrodes or smaller amounts of reactants result in correspondingly

---

[7] To emphasize the problem the partial reaction equations only are summarized; no simplification of stoichiometry has been applied.

shorter times of electrolysis. A temperature of about $T = 17\,^{\circ}C$ should be maintained during electrolysis. Carbon dioxide is passed through the solution continuously. An optimum flow is obtained when the solution is colored amber-yellow. Formation of the product can be observed after a short time of electrolyis, when small particles of yellow-colored iodoform float on the solution surface. After the end of the electrolysis they are collected with a filter, washed with water and dried. To calculate the yield with respect to the starting amount of KI the stoichiometrically correct simplified version of Eq. (7.13) is used:

$$3\ I^- + CH_3CH_2OH + 9\ OH^- \rightarrow CHI_3 + CO_3^{2-} + 7\ H_2O + 10\ e^- \qquad (7.15)$$

In the present example the yield is about 50%.

Experiment 7.10: Electrochemical Production of Potassium Peroxodisulfate

Task
Potassium peroxodisulfate is synthesized by electrochemical oxidation of sulfate ions.

Fundamentals
Anodic oxidation of hydrogen sulfate ions yields peroxodisulfuric acid according to

$$2\ HSO_4^- \rightarrow H_2S_2O_8 + 2\ e^- \qquad (7.16)$$

which precipitates as the respective salt when a sufficient cation concentration is present. A reaction path via has been proposed:

$$2\ HSO_4^- \rightarrow SO_4^- + e^- \qquad (7.17)$$

Recombination of the intermediate yields

$$2\ SO_4^- \rightarrow S_2O_8^{2-} \qquad (7.18)$$

From the respective salt, hydrogen peroxide can be obtained, but this process is not competitive when compared with, e.g., the anthraquinone process.

Execution
Chemicals and instruments
Saturated aqueous solution of $KHSO_4$
Platinum wire electrode (anode)
Platinum or nickel sheet electrode (cathode)
Beaker
Glass tube closed at the bottom with a porous frit or a cotton plug
Large beaker with cooling ice-water mixture
Power supply

Setup

The small beaker filled with the solution of $KHSO_4$ is placed in the cooling bath. The anode is placed in the solution. The cathode is placed inside the glass tube mounted with its bottom close to the bottom of the beaker.

Procedure

At a voltage of about $U=12$ V and a current of about $I=1.5$ A about half an hour of electrolysis is performed. The electrical power applied to the cell is in part converted into heat (Joule heating), so it coud be necessary to add further ice. After some time white crystals of $K_2S_2O_8$ precipitate. They can be collected on a filter and washed with a small amount of cold water. Identification is possible by, e.g., oxidation of iodine or Mn^{2+} ions.

Question

What by-product is generated at the cathode?

Experiment 7.11:
Yield of Chlor-alkali Electrolysis According to the Diaphragm Process

Task

In chlor-alkali electrolysis using the diaphragm process Faradaic yield is determined by titration of the generated sodium hydroxide solution.

Fundamentals

In chlor-alkali electrolysis using the diaphragm process (EC: 404), the electrolyte solutions in the anode and the cathode compartment are separated by a porous diaphragm, whose task is to avoid mixing of the produced gases (hydrogen and chlorine) resulting in a highly reactive mixture and mixing of the electrolyte solutions. Hydroxyl ions generated at the cathode and transported to the anode could react with the generated chlorine, yielding undesirable by-products and lowering the yield. Initially, diaphragms were prepared from mortar mixed with salt; leaching out the salt resulted in a highly porous diaphragm. This material was later replaced by asbestos fiber. These diaphragms are not perfect, as even after optimization of direction and speed of solution flow inside the cell, the undesired transfer of hydroxyl ions could not be suppressed completely. This drawback of the diaphragm process becomes obvious in measurements of the Faradaic yield of the electrolysis calculated with respect to the amount of electrical charge consumed. The obtained masses of chlorine, hydrogen, and caustic are smaller than expected.

Execution
Chemicals and instruments
Aqueous solution of NaCl 20 wt%
Absorption bottle with caustic solution for absorption of chlorine
Phenolphthalein indicator solution
Aqueous solution of HCl 0.1 **M** for titration
Pipet 5 ml
Electrolysis cell
Power supply DC 6 A
Watch

Setup
The cell used is depicted in Fig. 7.8.

A graphite rod (central electrode of a Leclanché cell, arc carbon) is pushed into one of two holes in a rubber stopper. Into the other hole a glass tube is pushed, and this is connected via a rubber tube to the absorption bottle used for chlorine absorption in an alkaline solution. The rubber plug is pressed into the top opening of a porous vessel (e.g. porous fired clay diaphragm, a glass tube with a porous bottom). This vessel is placed in the middle of the beaker serving as the electrolysis cell. As the cathode, a steel plate with many holes or a spirally wound iron wire is placed close[8] to the wall of the beaker. A bent iron wire placed between diaphragm and cathode is used as a stirrer.

Procedure
Electrolysis is performed at a constant current of $I=2$ A. After ten minutes the first sample of caustic solution is retrieved from the cathode chamber after thoroughly mixing this solution with the stirrer. The sample is titrated to determine the content of NaOH. This determination is repeated three times. When calcu-

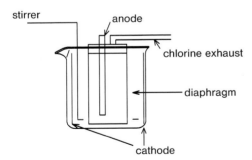

Fig. 7.8 Electrolysis cell for the diaphragm process.

8) The cathode should not be placed too close to the wall in order to avoid undesired concentration gradients between inner and outer surface. The holes in the sheet serve the same purpose. If these precautions are not taken substantial measurement errors may be observed.

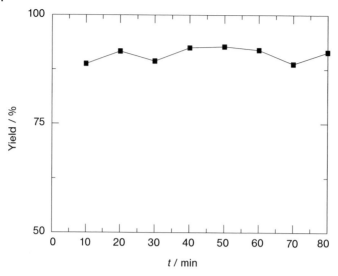

Fig. 7.9 Faradaic yield of produced NaOH during electrolysis using the diaphragm process.

lating the yield based on current and passed time the decrease in the solution volume in the cathode chamber must be taken into account.

Evaluation
A typical result is shown in Fig. 7.9. As expected, because of the imperfect effectiveness of the diaphragm, the yield decreases slowly as a function of time of operation. Incomplete mixing by the stirrer and imperfect precision during sample collection result in some data scatter.

Question
During extended electrolysis the amount of carbon dioxide in the chlorine grows. What side reaction is the source?

Literature
D. Pletcher and F.C. Walsh: Industrial electrochemistry, Blackie Academic & Professional, London 1993.

Appendix

The calibration of analog-to-digital converters (AD-converters) embedded in plugin-cards for computers requires a high-precision voltage source. Frequently employed standard cells like, e.g., the Weston cell provide voltages too low for this purpose; the voltage should be as close as possible to the maximum input voltage of the converter in order to allow a precise adjustment of the gain (slope) of the converter. A circuit sufficient for most applications based on generally available components and easy to build is depicted below (Fig. A.1). Initial adjustment of the circuit requires access to a precise voltmeter.

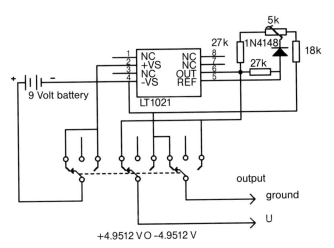

Fig. A.1 Circuit diagram for a reference voltage source supplying ± 4.9512 V.

Experimental Electrochemistry. A Laboratory Textbook. Rudolf Holze
Copyright © 2009 WILEY-VCH Verlag GmbH & Co. KGaA, Weinheim
ISBN: 978-3-527-31098-2

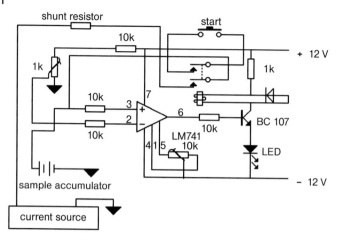

Fig. A.2 Circuit diagram for a cell voltage supervisor device.

Control of the discharge of an accumulator needs a comparator circuit shutting off the discharge current once a predefined lower voltage level has been passed. This can be performed easily with the following circuit (Fig. A.2). Discharging is started by pressing the "start" button, the ongoing process is indicated with the LED. Once the predefined voltage has been passed the discharging current is interrupted; discharging will resume only after pressing the "start" button again.

Index

a

Abrasive Stripping Voltammetry   183
Acetoacetic Ester   227
Acetyl Acetone   223
Activation   143, 145, 154, 159
activation potential   84
Activities of solid phases   16
Activity coefficient   17
Adsorbate geometry   200
Aeration cell   139
Agar   45
Aliphatic nitro compound   54
Alkaline ester saponification   60
Amperometry   168
Amplification method   186
Aniline   113, 197
Anodic oxidation of an aluminum   220
Anodic stripping voltammetry   180
Aromatic Hydrocarbons   110
Arrhenius equation   59
Arrhenius plot   59
Autocatalysis   33
Autocatalytic reaction   32

b

Berlin blue   204
Berlin yellow   204
Biamperometry   167
2,3-bis-ethoxycarbonyl-succinic acid diethyl ester   226
Bode plot   137
Breakthrough potential   84
Bromatometric titration   168
Bromination   229
Bromoform   229
Butler-Volmer equation   74

c

Calibration curve   21
Cell constant   48
Cell without transfer   15
Cells without diffusion potential   22
Cementation reaction   217
Cerimetry   26, 28
Charge transfer reaction   88
Charge-discharge yield   208
Chemical Constitution and Electrolytic Conductance   54
Chemisorbed oxygen   82
Chronoamperometry   122, 124
Chronocoulometry   124
Clark cell   7
Collection efficiency   131
Compliance of the potentiostat   106
Concentration cell   32, 141
Concentration gradients   88
Conductance measurement cell   47
Conductance Titration   52
Conductometrically Indicated Titration   161
Constitution isomerism   54
Contact corrosion   137
Convection   119
Corrosion cells   137
Corrosion protection   114
Cottrell equation   122, 124
Coulometric Titration   166
Coupled homogeneous or heterogeneous chemical reactions   99
Current follower   7, 10
Current-potential curves   208
Cyclic Voltammetry   77
Cyclic Voltammetry of Organic Molecules   99
Cyclic Voltammetry with Microelectrodes   95
Cyclic voltammograms   78

Experimental Electrochemistry. A Laboratory Textbook. Rudolf Holze
Copyright © 2009 WILEY-VCH Verlag GmbH & Co. KGaA, Weinheim
ISBN: 978-3-527-31098-2

d

Daniell element   12
Data Recording   10
DC-polarography   178
Debye-Hückel region   22
Debye-Hückel theory   17
Decamethyl ferrocene   6
Decaphenyl ferrocene   6
Decoloration   33
Decomposition voltage   36
Degree of dissociation   48
Dehydration of formaldehyde   69
Deionized water   3
3,4-diacetyl-hexane-2,5-dione   224
Diaphragm process   234–235
Differential aeration cells   140
Differential double-layer capacitance   189
Differential Potentiometric Titration   28
Differential pulse polarography   179
Diffusion potentials   6
N,N-dimethylaniline   99
6,10-diphenylanthracene   110
Dissociation constant   48
Double layer region   80
Double-layer capacity   83
Double-layer charging   82
Double-layer region   83
Drift speed   63
Drift velocity   44

e

Effective ion radius   68
Electro-optic properties   196
Electrochemical cells   8
Electrochemical impedance measurements   8
Electrochemical roughening   200
Electrochemical series   11, 14
Electrochromism   114, 203
Electrode function   22
Electrode Impedances   134
Electrodes of the second kind   15
Electrogravimetry   163
Electrophoresis   44
Electroreduction of Formaldehyde   69
Emeraldine   116
Ester Saponification   59
Ethyl acetate   59
Eudiometer   58
Exchange current density   76, 88

f

Faradaic reactions   80
Faradaic yield   208
Faraday's Law   56, 151, 166
Faraday's Second Law   57
Ferrocene   6, 105
Ferroxyl indicator   142
Fick's first law   119
Fick's second law   119
First Faraday Law   57
Flade potential   84
Formal potential   106
Formic acid   80
Fuel Cell   213

g

Galvanostatic coulometry   166
Galvanostatic Measurement   74
Galvanostatic Step Measurements   118
Gel electrophoresis   44
Gel-like electrolyte   44
Gibbs energy   12, 38
Gibbs equation   13
Glass electrode   22
Grotthus jump mechanism   63

h

H-cell   8
Half-cell   12
Haloform reaction   229
Hard solder   4
High input impedance voltmeter   12
High-precision voltage reference   10
Hittorf Transport Number   63
Homogeneous oxidation reaction   32
Hydrogen electrode   5

i

Ilkovič equation   179
In situ investigation   195
Indirect Anodic Dimerization   227
Indium-doped tin oxide   196
Infrared spectroelectrochemistry   201
Intraband transition   196
Iodination   231
Iodoform   232
Ion sensitive electrode   152
Ionic mobility   44
Ionic radius   63
ISE   152
ITO-electrode   195

j

Joule heating   225, 229

k

Kinetic currents   69
Kinetics of the Oxidation   32
Kipp apparatus   218
Kohlrausch's square root law   48
Kolbe Electrolysis   222

l

Lead acid accumulator   205
Linear diffusion   95
Local cell   137
Local corrosion elements   142

m

Macro corrosion element   137
Malonic Acid Diethylester   226
Mass transfer   88
Mean activity coefficient   15, 18, 20
Mechanical perturbation   143
N-methylaniline   115
Microelectrode   4
Microelectrodes   95, 97
Migration   118
Mobility   8, 15, 44–46, 63, 68
Molecular electronics   114

n

Nernst equation   11–12
Nickel-Cadmium Accumulators   210
Nigraniline   116
Nitroethane   54
Normal hydrogen electrode NHE   6
Numerical Simulation of Cyclic
      Voltammograms   93

o

Optical absorption   196
Optically transparent electrode   196
Oscillating reactions   147
Ostwald's dilution law   48
OTE   196
Overoxidation   117
Overvoltage   40
Oxalic Acid   32
Oxyhydrogen coulometer   66

p

Palladium/gold wire   5
Paper electrophoresis   46–47
Partial current densities   74

Passivation   143, 145, 154, 159
pH-Measurements   21
Planar diffusion   95
Platinum black   5
Polarization   36
Polarographic analysis of anions   185
Polarography   174
Polyacrylamide gel   44
Polyaniline   113, 197
Polypyrrole   114
Polythiophene   114
Porous electrodes   206
Potassium peroxodisulfate   233
Potentiometrically Indicated Titrations   21
Preaccumulation   180
Precipitation titration   161
Prussian blue   128, 203
PtO-chemisorption layers   83
Purification   3

q

Quinhydrone electrode   22

r

Ragone diagrams   211
Randles-Ševčik equation   180
Raney metals   213
RCL-bridge   49
RDE   126
Reaction enthalpy   38
Reaction entropies   38
Reaction entropy   13
Redox Titrations   26
Reference electrode   4
Relative hydrogen electrode RHE   6
Rotating disc electrode   126
Rotating Ring-Disc Electrode   131
Roughness factor   82
RRDE   131

s

Sacrificial anode   137
Salt bridge   9, 13
Salt bridges   8
Salt Water Drop Experiment   142
Sampling polarography   179
Sand equation   119
SERS   199
Shunt resistor   10
Silver chloride electrode   4
Slow Scan Cyclic Voltammetry   85
Soft solder   4
Solid poymer electrolyte fuel cell   214

Solvated ions   47
Specific conductance   48
Spectroelectrochemistry   195
Spherical diffusion   95
Stainless steel   145–146
Standard electrode potentials   15
Standard hydrogen electrode   11
Standard hydrogen electrode SHE   6
Standard Potential   18
Standard potentials   11
Stationary phase   46
Stia counters   57
Stokes friction   63
Styrene   168, 173
Surface enhanced Raman spectroscopy
   199
Symmetry coefficient   88, 92

t
t-butanol   190
Tafel equation   75
Tafel plot   76, 128, 130
Tafel plots   129, 206
Tafel slope   75, 76

Temperature coefficient   39
Tensammetry   188
N,N,N′,N′-tetramethyl benzidine   102
Throwing power   218
o-toluidine   115, 197
Tool steel   145–146
transference number   63
Transition times   118
Triangular voltage sweep method   78
Turnbull's blue   142
turning point   33

u
Ultrapure water   3
Unpolarizable reference electrode   171
UV-Vis spectroscopy   195

v
Voltmeters   7

w
Weak acids   25
Weston cell   7